Conservation of Threatened Desert Plants

About the Authors

Dr. Suresh Kumar, an M.Sc. in Botany in 1977 from Univer sity of Delhi, Delhi was honoured with professor P. Maheshwari Memorial Award for achieving first position. He earned his PhD on vegetation ecology of Jaisalmer district in 1988 from Jai Narain Vyas University, Jodhpur. Since 1978 he has served at CAZRI Jodhpur. He has been Sr. Scientist (1986-1998) and Principal Scientist (1998-2009) and is currently Head, Division of Integrated Land Use Management and Farming System, CAZRI, Jodhpur. He has headed research programmes funded by ACIAR, JICA, MoEF, DST, DBT, NMPB, RSMML, BLMCL, NRSA, SAC. He has surveyed, documented and mapped vegetation of almost all districts of western Rajasthan using satellite data and ground truthing. This extensive knowledge about vegetation and its management became evident in successful completion of rehabilitation of sand dunes (through aerial seeding), rocky habitats, limestone, gypsum and lignite mine spoils. Conservation of threatened species in all these rehabilitations was successfully demonstrated. He has 246 publications, of which 64 are in peer reviewed journals and 9 research monographs/books. He has been invited to deliver lectures in many national institutions/universities and international institutions in Japan, Australia, Sharjah and USA. He has also visited Cambridge, UK. He has been honoured with ICAR coveted Team Research Award in 1984 and Dr. K. A. Shankarnarayan Award in 1992. He has been member of important national committees of MoEF, New Delhi, such as Expert Group on Botanical Gardens, Expert group on Desert Biosphere Reserve and National Biodiversity Strategy and Action Plan. He has recently been nominated to International Cactus Committee to represent South Asia. The Desert Botanical Garden under his supervision since 1996 has been conferred the status of a "Lead Garden" by MoEF, New Delhi for excellent conservation efforts. He is currently handling a DBT programme on All India Network Project on "Preventing extinction and improving conservation status of threatened plants through application of biotechnological tools".

Dr. Chandan Singh Purohit obtained his M. Sc. Degree in Botany from MDS University, Ajmer in 2002 and M.Phil. degree for his work on "Taxonomic description and Nutritional value of some desert plants of Bikaner District of Rajasthan" from Maharaja Ganga Singh University, Bikaner in 2004 and his Ph.D. degree for his work on "Study of the Grasses of North-west Rajasthan" in 2009 from Maharaja Ganga Singh University, Bikaner. He has expertise in identification of Poaceae, conservation and population status study of RET & Endemic plants of Rajasthan. Presently he is working at Botanical Survey of India (BSI), Sikkim Himalayan Regional Centre, Gangtok, Sikkim and involved in study of Flora of Singhba Rhododendron Sanctuary and Phenology of Endangered, Endemic and threatened plants of Sikkim.

He has done extensive field work in Indian desert and most part of Rajasthan for exploration, population studies and conservation status of Poaceae and other RET & Endemic plants. His acute sense of observations related to taxonomy, nomenclature, distribution, phyto-geography and conservation status in respect of above plants brought out results that were published in various journals. He has published one Book entitled 'Grasses of North-west Rajasthan'. He has so far published 33 research papers, 3 extension folders, 1 book chapter, 12 research articles and 9 technical reports.

Conservation of Threatened Desert Plants

Suresh Kumar
Chandan Singh Purohit

SCIENTIFIC
PUBLISHERS (INDIA)

Published by:

Scientific Publishers (India)
5 A, New Pali Road, P.O. Box 91
Jodhpur 342 001 (India)

E-mail: info@scientificpub.com
Website: www.scientificpub.com

Branch Office
Scientific Publishers (India)
4806/24, Ansari Road,
Daryaganj
New Delhi - 110 002 (India)

ISBN: 978-81-7233-878-7

Cover Photo - Flower of *Moringa concanensis* (Photo by S. Kumar)

Typeset by Rajesh Ojha
Printed in India

Dedicated to
All taxonomists and field ecologists
who have toiled in this desert to
document floristic wealth

फोन / Phone : (0091) 033-2321 9334/5775
(0091) 033-2321 3891/3848

भारतीय वनस्पति सर्वेक्षण
निदेशक का कार्यालय,
सी.जी.ओ. कॉम्प्लेक्स, तृतीय एम.एस.ओ. भवन,
ब्लॉक-एफ (पाँचवां तल), डी.एफ. ब्लॉक,
सेक्टर 1, साल्ट लेक सिटी, कोलकाता - 700 064

भारत सरकार
GOVERNMENT OF INDIA
पर्यावरण एवं वन मंत्रालय
MINISTRY OF ENVIRONMENT & FORESTS

भारतीय वनस्पति सर्वेक्षण
BOTANICAL SURVEY OF INDIA

फैक्स/Fax : (0091) 033-2321 5631
ई-मेल/E-mail : bsi_headquarter@rediffmail.com

BOTANICAL SURVEY OF INDIA
Office of the Director
CGO Complex, 3rd MSO Building,
Block F (5th Floor), DF Block,
Sector 1, Salt Lake City, Kolkata - 700 064

संख्या / No./BSI 8/1/PA/2012

दिनांक / Date 13.04.2012

FOREWORD

Conservation of Biodiversity in all bio-geographic regions of our country has been a prime mandate of Ministry of Environment and Forests (MoEF), New Delhi since beginning through Botanical Survey of India (BSI), Zoological Survey of India (ZSI), Forest departments, Universities and Research Institutes. However, this activity became more focused after the adoption of Convention on Biological Diversity (CBD). Accordingly action plans were made to restore, maintain or reduce the decline of population of species of selected taxonomic groups. Efforts have also been underway to improve the status of threatened species. The Lead Garden Scheme of the Conservation and Survey Division of Ministry of Environment and Forests, Government of India with BSI as coordinator has aimed to achieve these goals in different regions of our country. One such important bio-geogrpahic region is Indian desert having 6.4% endemic species and nearly 50 species threatened in different categories. The responsibility of improving their status by *ex-situ* conservation of some 20 species was assigned to the Desert Botanical Garden at Central Arid Zone Research Institute (CAZRI), Jodhpur some three years ago. This garden has been re-designated as Lead Botanical Garden in the Indian Arid zone. Beside providing financial support for infrastructural development, the BSI has also been associated in overall supervision of this programme. The authors have investigated these species and finalized their conservation protocols. The information provided covers not only taxonomic characterization but also gives an account of threats faced by each species and conservation effort made for each species at the Lead Garden in CAZRI. The presentation includes photographs and the lucid description that can be easily understood by a conservation enthusiast as well as researcher, alike. I hope that the useful information provided on multiplication of each species will be of much use by officials of forest department, universities and other institutions engaged in conservation of threatened species. I congratulate the authors Dr. Suresh Kumar and Dr. Chandan Singh Purohit for bringing out such a compendium of information which will go a long way in the conservation history of threatened species in India.

Dr. P. Singh
Director, Botanical Survey of India
Howrah, Calcutta

PREFACE

Conservation of plants has been continuing since time inmemorial in view of its importance in saving the environment and humanity as a whole. However, last century has seen a very rapid degradation of environment resulting in wiping out of many species from the face of the earth and pushing many more to the brink of disappearance and vulnerability. While uses of many species are now getting known, it is imperative that we can't afford to make species vulnerable to such threats until we discover their economic importance. Hence, all countries have started conservation programmes, more so after adoption of Convention on Biological Diversity (CBD). In India, Ministry of Environment and Forest, New Delhi (MoEF) is nodal ministry of Government of India to implement provisions of CBD and it is in that context all bio-geographic regions of country have been taken up to conserve flora of each region in the respective Lead Botanical Gardens at a specified place. The Desert Botanical Garden at Jodhpur was given this responsibility and re-designated as Lead Garden of Indian desert in 2008. Some 20 threatened plant species were taken up in this progarmme for conservation at Lead Botanical Garden at Jodhpur. This book gives a vivid account of our efforts as well as over achievements in that direction.

The introductory chapter gives a historical account of CBD, its targets, implementations by various governments as well as the steps taken by the MoEF in this direction. The next chapter gives an overview of environment, flora and vegetation of arid region in India. Description of different types of threats to biodiversity of Indian desert constitutes another chapter. The next 20 chapters are devoted to 20 species, of which *ex-situ* conservation in the Desert Lead garden has been attempted. Each chapter has information on taxonomic citation, physiognomic description, taxonomic description, uses of these species and threats faced by it. The presentation is in the question answer style to increase the readability so as to get across the wide spectrum of conservation fraternity. In each chapter, it has also been described as to what efforts have been made to conserve a species at CAZRI. These chapters have been provided with large number of coloured photographs to make it as pictorial as possible. The bibliography concludes the book. Authors would welcome any suggestions for its improvement in future.

Authors

ACKNOWLEDGEMENTS

We are immensely grateful to Dr. M.M. Roy Director, Central Arid Zone Research Institute (CAZRI), Jodhpur (Rajasthan) and Dr. K.P.R. Vittal, Former Director, CAZRI, Jodhpur for their guidance and encouragement for carrying out this work. We are also thankful to the Ministry of Environment and Forests (MoEF), New Delhi for kindly approving the proposal and timely financial support. We are indebted to Dr. P. Singh, Director, Botanical Survey of India, Kolkata for kindly going through the manuscript and writing its 'Foreword'. We are also grateful to Dr. R.P. Pandey, Deputy Director, Arid Zone Regional Centre, Botanical Survey of India, Jodhpur for suggesting some improvements. We express our gratitude to the authorities of various institutions viz. Botanical Survey of India, Jodhpur; Botanical Survey of India, Pune; St. Xaviers College, Mumbai; Central Herbarium, Forest Research Institute, Dehradun and Herbarium at CAZRI, Jodhpur for the facilities provided to work in their herbaria.

We are obliged to Dr. G.V. Subrahmanyam, Advisor and Dr. R.K. Rai, Director, CS Division, MoEF, New Delhi for their guidance during reviews as well as timely financial support. We also thank Chief Administrative Officer and Finance & Account Officer and their staff at CAZRI who provided the necessary administrative and financial support. Dr. Farzana Parveen, SRA, Dr. Manish Mathur (T-5), Sh. Narayan Ram (T-5), Abhay Singh (T-5), S.K. Sankhla (T-5), Sh. V.K. Harsh (T-4), B.S. Choudhary (T-4), S.R. Bhaker (T-

4), C.S. Choudhary (T-3), Sh. Ravikiran Kulloli (SRF), Miss Deepika Acharya and Miss. Khushbu Rawat (SRF) deserve our special appreciation and thanks for their untiring assistance all the time during this period. Continuous and sincere participation of the staff of Desert Lead Botanical Garden at CAZRI at all times in all the activities of this programme is gratefully acknowledged.

Authors

CONTENT

1 INTRODUCTION

Biodiversity is the variety of life on earth. It comprises the variability within species, among species and of ecosystem. Biodiversity provides essential ecosystem services and is basis of ecosystem functioning and life support. In view of this, all nations came together between 1989 to 1992 to negotiate "The Convention on Biological Diversity" (CBD) under the auspices of the United Nations Environmental Programme. The treaty was opened for signature during the United Nations Conference on Environment and Development in Rio de Janeiro in June 1992 and came into force on 29th December, 1993. May 22 each year is celebrated as International Biodiversity Day. CBD has following three objectives:

— The conservation of biological diversity

— The sustainable use of its components and

— The suitable benefits arising out of the utilization of genetic resources.

The 188 member countries and European Union which are signatories to CBD, including India, follow global legal framework for action on biodiversity. Accordingly seven thematic work programmes have been developed by the Parties under CBD, of which biological diversity of dry and subhumid lands is one of the programmes. In each of these programmes, 10 goals and targets were set to be achieved by the year 2010. In the first goal on protecting components of biodiversity, item 2 relates to promoting conservation of species diversity. This has following two targets:

1. Restore, maintain or reduce the decline of populations of species of selected taxonomic groups
2. Status of threatened species improved (CBD website online, 2011).

In order to implement the commitment by Government of India to CBD, Ministry of Environment and Forests (MoEF) launched a major programme of financial support to Botanical Garden across the country for *ex-situ* conservation of rare, endangered, threatened and endemic (RET & E) species in all major bioclimatic zones. In that direction, one or two prominent Botanical Gardens have been selected and designated as Lead Botanical Garden, in each of the major biomes in India. The botanical garden at Central Arid Zone Research Institute, Jodhpur (CAZRI), has been designated as Lead Garden of Desert Biome. Some 20 RET&E species have been assigned to Desert Botanical Garden for their *ex-situ* conservation. Efforts have been made in the last three years to accomplish the *ex-situ* conservation of these species. Details of these species, conservation efforts, threat to their survival and role of public in conserving it has been described in this publication. Authors hope that the information provided herein will be useful to conservation enthusiasts, general public, researchers, foresters, policy makers and various government and non government agencies.

We shall feel amply rewarded if this book could further the cause of resource conservation, maintenance of genetic, structural, spatial and functional diversity that will go a long way in enhancing livelihood support for the poor people of the Indian Desert.

2 ENVIRONMENT, FLORA AND VEGETATION

The Indian hot arid zone (31.7 m ha) has peculiarities of climate, habitats and demographic trends. It experiences extremes of temperatures (50°C in summer to -4°C in winter), low and erratic rainfall (100-420 mm/year) with high coefficient of variation, high wind speed (30-40 km/hr), high evapo-transpiration (1500-2000 mm/year), light textured soil having poor fertility and low water retention capacity and scarce water resources, brackish groundwater and declining water table. Of the sandy, rocky, saline and alluvial habitats, 30% are wastelands. Nearly 58% area of western Rajasthan is under different types of sand dunes. Demographically, it is the most thickly populated (108 people/km^2) desert of the world. From 1961 to 2011, human population has increased by 194% and livestock density, by 72%. Thus, this is a climatically constrained low producing desert area having hazards of wind and water erosion, salinization, urbanization and mining, making it a fragile ecosystem (CAZRI, 2007).

Despite inhospitable and harsh climate, as well as anthropogenic pressures, Indian arid zone has 682 species belonging to 352 genera and 87 families. Some 63 species are introduced belonging to 37 genera and 9 families (Bhandari, 1990). This floristic wealth is contributed from various regions of the world: 37% species are African elements, 20.6% species are oriental, 14% are tropical and 10.3% are cosmopolitan. Compared to other deserts of the world, this desert has higher number of endemic species, i.e. 6.4%.

Poaceae and Leguminoseae are the largest families. Some 331 species of 44 families are ephemerals. Of these 52 species are winter annuals; 17 are winter and monsoon annuals and 282 species (38.7% of total flora) monsoonal, belonging to 184 genera and 44 families. Of these, maximum monsoonal ephemerals are from family Poaceae (50), followed by Fabaceae (25), Cyperaceae (20), Asteraceae (20), Euphorbiaceae (15), Convolvulaceae (13), Acanthaceae (11) and so on (Kumar, 2005). All these species have typical habitat-plant cover relationship resulting in major vegetation types on hills and rocks outcrops, piedmonts and pediments, alluvial plains, saline flats, fresh water lakes/talavs, river and stream banks, sandy hummocky plains and sand dunes of varying shapes and sizes (Kumar, 1998). Their grass cover type and woody perennial cover type under degraded and non degraded condition have different floristic elements (Table 1) (Kumar, 1994).

Even though vegetation in the desert is sparsely distributed, it is an important source of life support for the local people. Through ages, people here have learnt to use these local plants in a large variety of ways- such as vegetables (40 species), seeds (27 species), fruits (27 species), fibres (8 species), ropes (3 species), gums and resins (7 species), medicinal uses (131 species) and many provide fuel wood and forage (Saxena, 1979, 1981).

Besides, immense variability has also been found in pearl millet (1700 lines), Moth bean (600 lines), Mung bean (70 lines), Cow pea (55 lines), Clusterbean (3000 lines), Brassicas (14526 lines), Taramera (511 lines), Sesame (3283 lines), Ber (80 lines), Date palm (20 lines), Pomegranate (25 lines), *Aloe vera* (97 lines), *Cucumis callosus* (558 lines), *Citrullus lanatus* (193 lines), *Cucurbita melo* var. *momordica* (90 lines), Salad kakri (*Cucumis* species - one line) and undocumented variability in grasses, trees and shrubs (Kumar, 1999).

Thus, arid zone of India not only has diverse communities in a large variety of environmental situations, but these species also have immense variability at genetic level. And this is against the normally held belief that deserts

Table 1 Vegetation type of Indian arid zone under degraded and non-degraded conditions

Rainfall (mm)	Habitats	Grass cover type	Associated herbage species Non degraded sites	Associated herbage species degraded sites	Associated woody perennial		Area of occurrence
					Non- degraded sites	Degraded sites	
300- 400 A B	Sand clay Loam to clay Loam alluvium	*Dichanthium annulatum, Desmostachya bipinnata*	*Cynodon dactylon, Digitaria adscendens, Brachiaria ramose, Tetrapogon tenellus*	*Aristida mutabilis, A. funiculata, Goniogyna hirta, Zornia diphylla, Indigofera cordifolia*	*Acacia nilotica, Prosopis cineraria, Balanites aegyptiaca, Capparis decidua, Salvadora oleoides*	*Mimosa hamata, Cassia auriculata, Crotalaria burhia, Calotropis procera*	N. Shri Ganganagar, Nagaur, SE Sikar, pockets in Bali, Jalore, Sirohi
	Sandy loam Soil-alluvium	*Ochthochloa compressa, Dichanthium annulatum, Desmostachaya bipinnata*	*Eragrostis poaoides, Cymbopogon jwaracusa, Eremopogon foleolantus, Indigofera cordifolia*	*Aristida* sp., *Cenchrus biflorus, Tragus biflorus, Dactyloctenium sindicum*	*Prosopis cineraria, Salvadora oleoides, Capparis decidua*	*Calotropis procera, Cassia auriculata, Indigofera oblongifolia, Crotalaria burhia*	Nagaur, Jodhpur, Pali, Jalore, Sirohi
100-450	Saline clay loam	*Sporobolus marginatus, Dichanthium annulatum*	*Chloris virgata, Aleuropus lagopoaedes, Cressa creatica, Cynodon dactylon, Ochthochloa compressa, Cyperus* sp.	*Aeluropus lagopoades, Cressa cretica, Oropetium thromaeum*	*Salvadora oleoides, Acacia nilotica, Tamarix aphylla*	*Prosopis juliflora, Indigofera oblongifolia*	Saline low land in entire Indian arid zone
300-400	Sandy-sandy loam fossilized and stabilized dunes	*Saccharum spontaneum, Saccharum benghalese*	*Desmostachaya bipinnata, Cenchrus* sp.	*Aristida* sp., *Dactyloctenium* sp., *Indigofera cordifolia*	*Prosopis cineraria, Acacia jacquemontii*	*Sericostoma pauciflorum, Aerva pseudotomentosa, Crotalaria burhia*	Sikar, Nagaur, Churu, Jhunjhunun
250-350	Well-drained sandy alluvial soil	*Cenchrus ciliaris, Cenchrus setigerus*	*Ochthochloa compressa, Eragrostis* sp.	*C. biflorus, Aristida* sp.	*Prosopis cineraria, Acacia nilotica*	*Calotropis procera, Crotalaria burhia, Aerva pseudotomensa*	Churu, Jhunjhunun, Sikar, part of Nagaur, Jodhpur, Pali
Upto200	Loose sandy soil	*Lasiurus sindicus*	*Ochthochloa compressa, Indigofera cordifolia, Tribulus alatus*	*Aristida funiculata, Cenchrus biflorus, Mollugo cerviana*	*Prosopis cineraria, Tecomella undulata, Calligonum polygonoides, Z. nummularia*	*Crotalaria burhia, Dipterygium glaucum, Lycium barbarum*	Bikaner, Jodhpur, Jaisalmer, Barmer, west Jodhpur and west Nagaur
125-350	Stabilized sand dunes with reactivated surface	*L. sindicus, Panicum turgidum*	*Cymbopogon jwarancusa, Cyperus arenarius.*	*Aristida funiculata, Dactyloctenium sindicum*	*Prosopis cineraria, Calligonum polygonoides*	*Dipterygium glaucum, Lycium barbarum*	Barmer, Jaisalmer, Bikaner, Jodhpur, part of Jalore and Nagaur
150-300	Sandy gravelly plains	*Ochthochloa compressa, Dactyloctenium sindicum*	*Tribulus terrestris, Aristida* sp.	*Oropetium thomaeum, Aristida mutabilis, Tephrosia* sp.	*Prosopis cineraria, Salvadora oleoides, Acacia leucoploea, Capparis decidua*	*Barleria prionitis, Rhus mysorensis, Lycium barbarum*	Jaisalmer, Bikaner, Jodhpur
125-350	All types of soils except saline	*Aristida* sp., *Eragrostis* sp., *Cenchrus biflorus*	*Oropetium thomaeum*	*Oropetium thomaeum*	*Capparis decidua, Ziziphus* sp.	*Prosopis juliflora*	Jalore, Jaisalmer, Jodhpur, Barmer Sirohi

have poor diversity. In fact, deserts are new-found micro-hot spots of diversity that need to be given due priority and attention, as these are the plants going to be donor of genes for providing resistance to adverse effects of changes in future climate of our world (Kumar *et al.,* 2009). Deserts will also be affected by changes in climate. It has been estimated that due to climate change, the desert regions of India in this century may register a decline in rainfall in its northern part by up to 30%, an increase in its southern and eastern parts by up to 15% and a gradual rise in temperature by 4-5°C everywhere (CAZRI, 2007). The monsoon rains are likely to be abnormal. Cloud bursts and floods in south and south eastern Rajasthan and southern Gujarat and recurring droughts in arid Rajasthan will become more common. Such a scenario would result in high wind and water erosion. Dust storms with more dust loads will increase. Soil moisture status will experience a change. Coastal areas of arid Gujarat are likely to have sea level rise and spoilage of underground water with sea water ingress as well as loss of land. Quest for bringing more land under plough will enhance degradation and desertification. It is in this context that conservation of biodiversity in Parks, arboreta and Botanical Gardens assumes importance and priority so that precious germplasm is saved *ex-situ* before they are lost. We describe in the following pages our attempts in this direction in the Desert Botanical Garden.

3 THREATS TO BIODIVERSITY IN INDIAN DESERT

Biodiversity is wide variety of ecosystem and living organisms: animals, plants, their habitats and their genes; finally to the communities themselves that compose the living parts of the multifarious ecosystems of the world.

Salient features of India's Biodiversity

- India is one of the 12 centres of origin of cultivated plants. It has- 5 world heritage sites (Kaziranga National Park-Assam, Keoladeo Ghana National Park-Rajasthan, Manas Wildlife Sanctuary-Assam, Nanda Devi National Park-Uttar Pradesh, Sundarban National Park, West Bengal) and 12 Biosphere reserves (Nilgiri-Kerala, Nanda Devi- Uttranchal, Nokrerk- Meghalaya, Manas- Assam, Sunderbans-West Bengal, Gulf of Mannar- Tamilnadu, Great Nicobar-Andaman & Nicobar, Similipal- Orissa, Dibru Saikhowa-Assam, Dehang Dabang- Andhra Pradesh, Pachmarhi-Madhya Pradesh and Kanchanjunga-Sikkim) as well as 6 Ramsar wetlands. Among the protected areas India has- 88 National parks and 490 sanctuaries covering an area of 1.53 lakh sq. km (Singh, 2002).
- All these protected site are spread in 10 bio-geographic regions including- Trans-himalayan, Himalayan, Indian Desert, Semi-arid zone, Western

ghats, Deccan Peninsula, Gangetic Plains, North-east India and Islands and coasts. Habitat diversity in India includes 200 types of forests, 8 types of wetlands, 20 agro-ecozones and several types of coasts. These have such species having ethno-botanical uses. 7500 species are used in folk medicines, 3900 spp. are edible, 700 useful in material and cultural requirements, 525, in fiber and cordage, 400, are fodder, 300, as pesticides, 300, yield gums and dyes and 100 useful as incense and perfumes.

Threats to Biodiversity

Increase in human and livestock population has enhanced pressure on natural vegetation threatening biodiversity in three ways

(1) Habitat Disruption

(A) Habitat Fragmentation is reduction in habitat area to form small fragments through over-harvesting, construction of roads, construction of canals and other developmental activity i.e., land use for commercial and residential purpose etc.

(B) Habitat Alteration is replacement of existing vegetation for commercial exploitation such as use of fast growing tree species, rice, groundnut and cotton cultivation, invasion in wetland vegetations by species like *Typha angustata, Arundo donax, Eichhornia crassipes, Imperata cylindrica* and *Saccharum spontaneum*. Large water bodies due to seepage of IGNP have attracted many species which were not seen earlier in the desert. Consequently, pressure on un-command grazing land has increased many folds because of expansion of agriculture in canal command areas reducing pasturelands and silvopastoral activity

(C) Habitat Degradation: This has resulted from direct and indirect consequences of human activities such as desertification and salinization, urbanization and effluent disposal, land use changes, water flow disruption, depletion of ground water table and increased pollution.

(2) Over Exploitation

Excessive pressure reduces the time of recovery of species/vegetation. Changes in land use pattern result in depletion in forest cover. Excessive cutting, browsing and over grazing, reduce the population of some species declining their regeneration potential. Besides, large-scale commercial mining for gypsum, lime stones, fuller's earth, sandstones, dolomite and lignite is causing environmental problems due to poor disposal of mine overburden, failure to rehabilitate mined area and loss of local vegetation.

(3) Biological Invasion

Biological invasion is introduction of species into areas that they would never have reached on their own such as - *Prosopis juliflora, Lantana camera, Arundo donax,* mesophytic rats/birds. Examples include impact of Indira Gandhi Nahar on habitat loss of *Lasiurus sindicus*, an important fodder grass by invasion of *Arundo donax, Typha* species in stagnent water around canal system, intrusion of mesophytic rats, aves, mammals etc., displacement of desert flora and fauna and finally change in land use and livestock pattern.

Above factors make these species rarer and threatened. In order to understand the magnitude of threat faced by a species, International Union for Conservation of Nature (IUCN) has decided different endangerment categories as follows.

Nature of Endangerment categories

There are nine clearly defined categories into which every taxon in the world (excluding micro-organisms) can be classified (Figure 3.1). Complete definitions of the categories based on IUCN red list guidelines (IUCN redlist guideline, 2010 online) are as follow-

(1) **Extinct (EX)** = A taxon is extinct when there is no reasonable doubt that the last individual has died.

(2) **Extinct in the wild (EW)** = A taxon is `extinct' in the wild, when it is known only to survive in cultivation, in captivity or as a naturalized population (or populations) well outside the past range.

(3) **Threatened:** It includes three categories

- **Critically endangered (CR)** = A taxon is critically endangered when the best available evidence indicates that it meets any of the criteria A to E for critically endangered and it is therefore, considered to be facing an extremely high risk of extinction (50% in 5 years) in the wild.
- **Endangered (EN)** = Same as CR but facing a very high risk of extinction (20% in 20 years) in the wild.
- **Vulnerable (VU)** = Same as EN but facing a high risk of extinction (10% in 100 years) in the wild.

(4) **Not-Threatened:** It includes four categories

- **Near Threatened (NT)** = It does not qualify for any of the above but is close to qualifying for or is likely to qualify for threatened category in the near future.
- **Least concern (LC)** = It does not qualify for any of the above and is widespread and abundant taxa are included in this category
- **Data deficient (DD)** = A taxon is data deficient when there is inadequate information to make a direct, or indirect, assessment of its risk of extinction based on its distribution and/or population status
- **Not Evaluated (NE)** = A taxon is not evaluated when it has been initially proposed but not

discussed and not yet been evaluated against the criteria for any reason, including uncertainty about taxonomic or wild status.

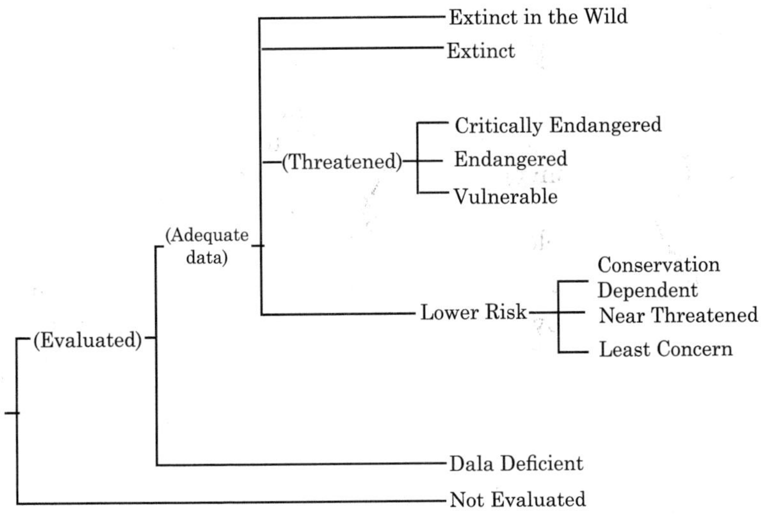

Fig. 3.1 Categories of Endangerment by IUCN Red list (Version 8.1, Aug. 2010)

The Criteria for Critically Endangered, Endangered and Vulnerable

A taxon is considered threatened if it meets any one of the criteria listed below. A taxon is assigned highest threat category as per the relevant criteria, if it meets 2 or more criteria or categories.

The five criteria are:

a. reduction in population size

b. reduction in extent of occurrence

c. reduction in area of occupancy

d. reduction in mature individuals

e. probability of extinction in the wild

Based on above, conservation of threatened plants is taken up either *ex-situ* or *in-situ*, depending upon species, resources, facilities and time available for this purpose.

(a) ***In-situ* Conservation:** This is the conservation of genetic resources through their maintenance with in natural or even human-made ecosystem in which they occur. This is an ideal system for genetic resources conservation. This type includes a system of protected areas of different categories, managed with different objectives to bring benefits to the society. National Parks, Sanctuaries, Nature Reserves, Natural Monuments, Cultural Landscapes, Biosphere Reserves etc. belong to this type of conservation.

(b) ***Ex-situ* Conservation:** This is conservation outside their habitats by perpetuating sample populations in genetic resource centres, zoos, botanical gardens, culture collections etc. or in the form of gene pools and gamete storage for fish; germplasm banks for seeds, pollen, semen, ova, cells etc. In this type of conservation seed banks, botanical gardens, pollen storage, tissue culture and genetic engineering have been playing important role.

Threatened species in the Indian Arid Zone: Above criteria have been used by Botanical Survey of India, Arid Zone Circle, Jodhpur, which has identified threatened plants of Indian desert (Pandey *et al.*, 1983) as follows:

S N.	Name of threatened plant species	Family
1	*Anogeissus sericea* var. *nummularia* King ex Duthie	Combretaceae
2	*Tecomella undulata* (Sm.) Seem.	Bignoniaceae
3	*Moringa concanensis* Nimmo ex Dalz. & Gibs.	Moringaceae
4	*Barleria prionitis* var. *dicantha* Blatt. & Halb.	Acanthaceae
5	*Commiphora wightii* (Arn.) Bhandari	Burseraceae
6	*Cullen plicata* (Delile) Stirton	Fabaceae

S N.	Name of threatened plant species	Family
7	*Melhania magnifolia* Blatt. & Hallb.	Sterculiaceae
8	*Pavonia arabica* var. *glutinosa* Blatt. & Hallb.	Malvaceae
9	*Pavonia arabica* var *massuriensis* Bhandari	Malvaceae
10	*Ziziphus truncata* Blatt. & Hallb.	Rhamnaceae
11	*Tephrosia falciformis* Ramaswamy	Fabaceae
12	*Seddera latifolia* Hochst. Ex Steud.	Convolvulaceae
13	*Acacia jacquemontii* Benth.	Mimosaceae
14	*Ammannia desertorum* Blatt. & Hallb.	Lythraceae
15	*Citrullus colocynthis* (L.) Schard.	Cucurbitaceae
16	*Ephedra ciliata* Fisch. & Mey. ex C.A. Meyer	Ephedraceae
17	*Ceropegia bulbosa* Roxb. var. *bulbosa*	Asclepiadaceae
18	*C. bulbosa* Roxb. var. *lushii* (Grah.) Hook.f.	Asclepiadaceae
19	*Neurada procumbens* L.	Rosaceae
20	*Lasiurus sindicus* Henr.	Poaceae
21	*Abutilon bidentatum* var. *major* (Blatt. & Hallb.) Bhandari	Malvaceae
22	*A. fruticosum* var. *chrysocarpa* Blatt. & Hallb.	Malvaceae
23	*Alysicarpus vaginalis* var. *venosa* (Blatt. & Hallb.) A. Pramanik & Toth.	Fabaceae
24	*Anticharis senegalensis* var. *caerulea* Benth. & Hallb.	Scrophulariaceae
25	*Aristida royleana* Trin. & Rupr.	Poaceae
26	*Cenchrus prieurii* var. *scabra* Bhandari	Poaceae
27	*Cenchrus rajasthanensis* Kanodia & Nanda	Poaceae
28	*Chloris quinquesetica* Bhide	Poaceae

S N.	Name of threatened plant species	Family
29	*Cleome gynandra* var. *nana* (Blatt. & Hallb.) Bhandari	Cleomaceae
30	*Convolvulus auricomus* var. *ferruginosus* Bhandari	Convolvulaceae
31	*Convolvulus blatterii* Bhandari	Convolvulaceae
32	*Digitaria pennata* var. *shettyana* Pandey	Poaceae
33	*Echinops rajasthanense* Pandey & Singh	Asteraceae
34	*Euphorbia jodhpurensis* Blatt. & Hallb.	Euphorbiaceae
35	*Farsetia macrantha* Blatt. & Hallb.	Brassicaceae
36	*Ipomoea carica* var. *semine-glabra* (Blatt. & Hallb.) Bhandari	Convolvulaceae
37	*Melhania futteyporensis* var. *major* (Blatt. & Hallb.) Santapau	Sterculiaceae
38	*Pulicaria rajputanae* Blatt. & Hallb.	Asteraceae
39	*Rosa clinophylla* Thory	Rosaceae
40	*Sida tiagii* Bhandari	Malvaceae
41	*Tribulus rajasthanensis* Bhandari & Sharma	Zygophyllaceae
42	*Odontanthera varians* (Stocks) Hk.f.	Asclepiadaceae
43	*Limeum indicum* Stock ex T. Andr.	Molluginaceae
44	*Urochondra setulosa* (Trin.) C.E. Hubb.	Poaceae

Ex-situ Conservation : Methodology in this study

From the above list, species listed at serial no. 1 to 20 were selected for *ex-situ* conservation in the Desert Botanical Garden. Following methodology was followed:

- Published literature, check lists, herbarium and other historical data were studied to understand past distribution and occurrence of these species in Indian arid zone and Rajasthan.

- Using satellite data, landscape maps were prepared to visit those habitats of probable occurrence.
- In a field visit, those locations of previously reported occurrence were searched for recording these species. Additionally entire habitat was also searched.
- After a species was located, its seeds/bulbs/cuttings were brought for regeneration/multiplication.
- Different botanical parameters and initial germination tests were carried out. Thereafter, detailed regeneration protocols of each species were standardised.
- These saplings were planted in the fields of Desert Botanical Garden to monitor damage due to abiotic and biotic factors as well as growth, so as to understand threats faced by these species in nature.
- All these aspects are detailed in following pages.

4 *ACACIA JACQUEMONTII* (MIMOSACEAE)

Acacia jacquemontii Benth. in Hook. Lond. Journ. Bot. 1: 499, 1842; Baker in Fl. Brit. India, 2: 293, 1878; Bhandari, Rev. Fl. Ind. Desert 132, 1990; Pandey in Shetty & Singh Fl. Rajasthan 1: 296, 1987; Singh & Singh, Biodiversity Desert National Park, 87, 2006.

Local name: Bu-banvali, Ratio-banwal.

Fig. 4.1 *Acacia jacquemontii* in the Indian desert

What is Bhu-banvalio or *Acacia jacquemontii*? Genus *Acacia* belongs to family Mimosaceae. It is a rare gregarious species and a good sand binder; often found on stable sand-dunes (Fig. 4.1). The plants remains apparently dry in the rainy season and develop leaves at end of winter period. The flower buds coming up at end of winter later on, flowers and fruits in summer. The flowers are somewhat sweet-scented.

How to identify it? A much branched, small, bushy shrub with stiff, smooth, brown branches; Leaves 2-pinnate, stipular spines straight, slender, 3 cm long, ivory-white, connate at base; pinnae 2-4 pairs, 6-12 mm long, common petiole 3 cm long, gland small, indistinct; leaflet 5-10 pairs, linear-oblong, glabrous; Flowers yellow, sweet scented; Inflorescence globose heads, 12-16 mm in diameter (Fig. 4.2); peduncles 2-3, slender, axillary, fascicled; bracts 2 inserted almost in the middle of peduncle; Calyx campanulate, the teeth short, deltoid; Corolla 3 mm long, lobes ovate-oblong, acute; Pods ovate-oblong, rounded at base, flat, straight, transversely reticulately veined, glabrous; Seeds 5-8, compressed (Fig. 4.3). Its flowering and fruiting occurs during March to May (Bhandari, 1990).

Fig. 4.2. Flowers and Buds of *A. jacquemontii*

Fig. 4.3. Seeds of *A. jacquemontii*

How does it differ from Banwal or *Acacia nilotica*? *Acacia jacquemontii* is much branched, small, bushy shrub having flat pods (Fig. 4.2) but *Acacia nilotica* is an evergreen medium sized tree having moniliform pods; pods are 0.9 to 2 cm wide (Fig. 4.4 & 4.5).

Fig. 4.4. Flower of *A. nilotica*

Fig. 4.5. *Acacia nilotica*

Where does it naturally occur? It is distributed in Pakistan, India [Gujarat, Punjab, Rajasthan, Delhi and Saurashtra]. In Rajasthan (Fig. 4.6), it is distributed in Bikaner, Jaisalmer, Jodhpur, Barmer, Nagaur, Churu, Jhunjhunun and Sikar.

Fig. 4.6. *Acacia jacquemontii*

What are its uses?

- The wood when burned gives out an intense heat and is, therefore, employed by gold, silver and iron-smiths. The bark of the root is used in the distillation of spirits. The leaves are threshed out and used as fodder. It also yields gum.
- Its bark is used to induce spontaneous abortion in women in any stage of pregnancy.

What are the threats to its survival?

Observations at the sites of its occurrence and during regeneration led to identify following threats to its survival:

- It is restricted to sand dunes and hummocks, where it occurs as the sand is stabilized.
- Predation of seeds by goats, camel and sheep, leaves very few seeds for population recruitment.
- It is over exploited plant because of its wood being used as fuel.
- Its fruits are eaten by cattlemen and wild animals.
- Its pods mature during summer (May-June) and rupture while attached to plants, scattering seeds on

sandy surface. These seeds are lost due to burial in sand, blown by wind or eaten by ants and small animals. This results in its very poor regeneration.

- Seeds are viable for short duration and if they do not find appropriate edapho-climatic conditions for germination, these are as good as lost.

What efforts have been made to conserve it? Localities of its previously reported occurrence were surveyed during this programme. It has been found from 22 sites in Bikaner (Hansera, Bamanwali, Lakhusar, Motigarh, Chhatergarh, Mahajan, Tejana, Karnisar, Kandeli, Bajju, Pholleji, Gusaisar etc.), 3 sites in Churu (Kanata, Baghsar, Ginari-badi), 3 sites in Jhunjhunun (Mintera, Near fort-Churu, Chophulla), 4 sites of Nagaur (Alay, Sri-Balaji, Laxminarayan Dhani, Kalteda), 1 site in Pali (Hirawas), 1 site in Jalore (Jagannath Goshala) and 4 sites in Sikar (Badi-bidodi, Dudwas, Hirawas-bada, Fatehpur).

- Sites of its occurrence were visited to collect seeds/ fruits.
- Seeds were collected from intact fruits before split.
- Seeds were shade dried and soaked in water in two ways for germination (1) 12 hours and (2) 24 hours (Fig. 4.7).
- Germination was good in 12 hours water soaking condition as compared to 24 hours.
- On emergence of plumule, seedlings were exposed to sunlight and transferred in pots with Farm Yard Manure (FYM) and soil mixture (1:2) and treated with fungicide (Bavistin) and termiticide (Foret).
- Growth in terms of leaf number, height, number of flower buds, fruits and seeds were periodically recorded till the end of growing period.
- Phenological events were also recorded simultaneously.
- Desert Botanical Garden has now a block of its plants (Fig. 4.8).

Fig. 4.7. Germination in seeds of *Acacia jacquemontii*

Fig. 4.8. A new block of *Acacia jacquemontii* developed in Desert Botanical Garden.

What can people do to protect it?

- Protection of existing plants in grazing lands and forests will promote its regeneration.
- It can be planted on field bunds and field boundaries as biofence.

5 *ANOGEISSUS SERICEA* VAR. *NUMMULARIA* (COMBRETACEAE)

Anogeissus sericea **Brand. var.** ***nummularia*** **King ex Duthie** Fl. Gangetic Plain 1: 340, 1903; Scott in Kew Bull. 33: 559, 1979; Pandey in Shetty & Singh. Fl. Rajasthan 1: 314, 1987; *A. coronata* Stapf in Bull. Misc. Inf. Kew 153, 1914; *A. rotundifolia* Blatt. & Hallb. In Journ. Bombay nat. Hist. Soc. 26: 525, 1919.

Local name: Dhauro, Indroka.

Fig. 5.1. Indroka at Kailana, Jodhpur

What is Indroka or *Anogeissus sericea* var. *nummularia*? *Anogeissus sericea* var. *nummularia* King ex. Duthie (Combretaceae) is endemic to Rajasthan. It is a rare (Fig. 5.1), but widely exploited multipurpose tree of the Indian desert. It has typical flower heads (Fig. 5.2).

How to identify it? A medium to large tree (Fig. 5.3); young branches and leaves rusty-silky pubescent or subtomentose; Leaves apex obtuse or obcordate, mucronulate or apiculate, silvery pubescent (Fig. 5.4); nerves 4-5, arcute; petioles 1-2 mm long; Flowers in axillary and terminal heads (Fig. 5.2), solitary or rarely branched, 3-4 mm long, rather thick; Calyx tube 3-4 mm long, wing dilated, yellow, pubescent, limb 5-dentate, 2-2.5 mm in diameter (Fig. 5.5); Fruits glabrous, wings cordate at the base. Its flowering and fruiting occurs during September to January (Bhandari, 1990).

Fig. 5.2. Flowers of *Anogeissus sericea*

Fig. 5.3. Bark of *A. sericea*

How does it differ from Dhaukra or *Anogeissus sericea* var. *sericea*? *Anogeissus sericea* var. *sericea* has elliptic leaf more than 2 cm long with silky hairs beneath; wings of its fruits jagged at the margins and its calyx cup deciduous but *Anogeissus sericea* var. *nummularia* has oval or obovate or orbicular leaf which is less than 2 cm long with short hairs beneath; wings of its fruits jagged at the apex only and its calyx cup semipersistent.

Fig. 5.4. Leaves of *A. sericea*

Fig. 5.5. Flower head of *A. sericea*

Where does it naturally occur? It is distributed in India (Rajasthan-Jhalawar, Jodhpur, Pali, Udaipur and Ajmer). It is a true out-breeder. It occurs in clumps at isolated places under protection on the margin of water bodies such as village talav or tank.

What are its uses?

- Wood is useful in making agriculture implements and furniture.
- Paste of leaves is applied on wounds for 3-5 days.
- Leaves are used as fodder.
- Charcoal is made from its wood.

What are the threats to its survival?

Observations at the sites of its occurrence and during regeneration led to identify following threats to its survival:

- The reason of rarity is its extensive exploitation for commercial purpose.
- Natural regeneration is through seeds, but the viability of the seeds is very low (0.1-0.2%) and vegetative propagation methods are not available for this species mainly because of the phenolic exudation from the cut ends, which inhibits rooting. As the rate

of regeneration and exploitation are not balanced, this species is rare in its natural stands (Nayar and Sastry, 1987). In addition, hostile environmental factors, excessive grazing and microclimatic changes threaten the survival of this species.

- Modification and destruction of the habitats has further declined its populations.

What efforts have been made to conserve it? Localities of its previously reported occurrence were surveyed. It has been found at village pond of Dhava village-Jodhpur, Kailana lake-Jodhpur and village Indroka on Nagaur road.

- Seeds were collected from wild plants from Jodhpur.
- Freshly collected seeds were shade dried and then germinated in partial shade.
- Seeds were sown in three ways (1) in polythene bags (2) directly in beds (3) seeds were first overnight soaked in water then sown in polythene bags (Fig. 5.6).

Fig. 5.6. Seed sown in polybags and their germination in Garden.

- Germination was higher in overnight soaked seeds.
- Young seedlings are eaten by birds and other animals. Hence, it was protected by wire mesh cages or covered by spinous plant twigs.

What can people do to protect it?

- If someone is removing its bark, he could be persuaded not to do it as it will kill the trees. It should be protected in farmland, 'orans', around habitations and in forests.
- It can be planted on field bunds and field boundaries. It provides fodder and fruits.

6 *BARLERIA PRIONITIS* SUBSP. *PRIONITIS* VAR. *DICANTHA* (ACANTHACEAE)

Barleria prionitis **L. subsp.** ***prionitis*** **var.** ***dicantha*** **Blatt. & Hallb** in J. Bombay Nat. Hist. Soc. 26: 811. 1919, non B. diacantha Nees 1847; Bhandari, Fl. Ind. Desert: 299. 1978; Pandey & Shetty in Shetty & Singh, Fl. Rajasthan: 635. 1991; Karthikeyan *et al.*, Fl. Pl. India – Dicotyl. 1: 7. 2009.

Local name: Procupine flower, Barleria, Vajradanti, Kundan

What is ***B. prionitis*** **L. subsp.** ***prionitis*** **var.** ***dicantha*?** The genus *Barleria* belongs to family Acanthaceae. The genus *Barleria* was dealt by Linnaeus (1753) on the basis of specimens collected from India. It is a pantropical but predominantly an Old World genus, with its main centre of species diversity in tropical East Africa, followed by South Africa and Asia (Balkwill & Balkwill, 1998). It is the third largest genus in the family Acanthaceae with c.300 species (Balkwill & Balkwill, 1997; Mabberley, 2008). Karthikeyan *et al.* (2009) enumerated 29 species, one subspecies and six varieties. Two varieties of *Barleria prionitis* are reported from Rajasthan i.e., *B. prionitis* var. *dicantha* and *B. prionitis* var. *prionitis*. The flowers of variety *dicantha* are produced profusely and the corolla is deciduous (Fig. 6.1).

Fig. 6.1. *Barleria prionitis* var. *dicantha*

How to identify it? It is a perennial shrub, erect, up to 1.5 m high, much branched, armed. Stems terete or 4-angled, glabrous; thorns simple, 8 or more per node, 1 – 3.5 cm long, whitish. Leaves ovate or elliptic, 6 – 15 × 1.5 – 4 cm, attenuate at base, ciliate at margins, cuspidate at apex, glabrous above, more or less pubescent beneath, especially on veins (Fig. 6.2); lateral veins 4 – 6 pairs; petioles 0.5 – 3 cm long, grooved above. Flowers sessile (Fig. 6.3), often solitary in lower axils, becoming spicate above, rarely in simple dichasial cymes, 5 – 7 cm long; bracts linear-lanceolate, 1.2 – 2 cm long, cuspidate at apex. Outer sepals linear-lanceolate, subequal, 1.5 –2.3 × 0.4 – 0.5 cm, cuspidate at apex, appressed-hairy inside; inner sepals linear-lanceolate, 1 – 1.5 cm long, cuspidate at apex. Corolla bilipped (4 / 1), pubescent outside, pale to golden yellow; tube 1.5 – 2 cm long, swollen, curved and glabrous at base; upper lobes oblong-obovate, 2–2.5 × 0.6 – 1.4 cm, rounded at base, entire at margins, acute at apex; lower one oblong to broadly ovate, deeply lobed. Fertile stamens 2, exserted; filaments 2–3 cm long, glandular-pubescent at base, yellowish; anthers c. 5 mm long, yellow; sterile stamens 2,1–2 mm long; anthers c. 1 mm long; staminode absent. Disc covering less than ¼ of ovary. Ovary ovoid, c. 4.2 mm, glabrous; style terete, 2.5–3.5 cm long, hirsute at base; stigma linear, c. 1 mm long, sticky, pinkish. Capsules ovoid, 1.5–2 × 0.5–0.8 cm, compressed, 2-seeded. Seeds ovoid, 7.5–9 × 4.5–7 mm, subcordate at base (Bhandari, 1990).

Fig. 6.2. Leaf and spine of *B. prionitis* var. *dicantha*

Fig. 6.3. Close up of flower of *B. prionitis* var. *dicantha*

How does it differ from *B. prionitis* L. subsp. *prionitis* var. *prionitis*? *Barleria prionitis* subsp. *prionitis* var. *dicnatha* has long corolla tube up to 6 cm long, lower lobe up to 3 cm long and total corolla length is 6 cm but variety *prionitis* has 4-4.5 cm long corolla and corolla tube is shorter (Table 1).

Table 1. Comparison of flowers in both varieties

	B. prionitis **L. subsp. *prionitis*** **var.** ***dicantha***	***B. prionitis*** **L. subsp. *prionitis*** **var.** ***prionitis***
Flowers-corolla tube length		

Where does it naturally occur? It is present in open dry rocky localities in deciduous forests. It is present only in India and Pakistan. In India, it is reported from Diu & Daman, Gujarat, Madhya Pradesh, Maharashtra, Punjab and Rajasthan. In Rajasthan, it is present only in western Rajasthan especially at Jodhpur and Barmer. Recent survey by CAZRI, Jodhpur revealed this plant at different rocky habitats of Nagaur and Pali.

What are it uses?

- It has numerous medicinal properties as it reportedly cures toothache, joint pains and has several cosmetic uses.
- A mouthwash made from root tissue is used to relieve toothache and cures bleeding gums.
- The leaves are used to promote healing of wounds and toothache. Because of its antiseptic properties, extracts of the plant are incorporated into herbal cosmetics and hair products to promote skin and scalp health.
- Its leaves are used to relieve pain. Its twigs are used as tooth brush.
- It is also used as ornamental plants.

What are the threats to its survival? Observations at the sites of its occurrence and during regeneration led to identify following threats to its survival:

- They have limited occupancy because of various reasons, including biotic factors.
- Modification and destruction of its habitats appear to be main cause of its less survival.
- Its aerial plant parts are affected by insect pests and fungal infections.
- It occurs on dry open rocky localities, where lack of moisture and suitable edaphic conditions further hamper its natural regeneration.
- Squirrels, rodents, birds and peacocks foraging its leaves, flowers and seeds also decline its populations resulting in low survival.

What efforts have been made to conserve it?

Localities were surveyed from where it has been reported by previous workers in four districts i.e. Barmer, Pali, Bhilwara and Tonk. Of these it has been found only at four sites i.e. Kalteda in Nagaur district and Bijapur, Sabalpura, Ranakpur in Pali district.

- Some 50 stem cuttings were brought along with local soil to plant it in the nursery of the Desert Botanical Garden at CAZRI, Jodhpur.
- Growth in terms of leaf number, height, number of flower buds, fruits and seeds were periodically recorded till the end of growing period.
- Fruits and seeds were collected and stored (Fig. 6.4, 6.5 & 6.6).
- Phenological events were also recorded simultaneously.

Fig. 6.4. Mature fruit of *B. prionitis* var. *dicantha* in garden

Fig. 6.5. Close up of fruit of *B. prionitis* var. *dicantha*

Fig. 6.6. Close up of seed of *B. prionitis* var. *dicantha*

- A new block of this plant is developed in Desert Botanical Garden, CAZRI, Jodhpur (Fig. 6.7 & 6.8).

Fig. 6.7. A new block of *Barleria* developed in Desert Botanical Garden

Fig. 6.8. One year old plants of *Barleria prionitis*

What can people do to protect it?

- Protection of existing plants in farmland, orans, around habitations and in forests could promote its regeneration.
- It can be planted on field bunds and field boundaries to serve as biofence.

7 *CALLIGONUM POLYGONOIDES* (POLYGONACEAE)

Calligonum polygonoides L. Sp. Pl. 530, 1753; Hook.f. Fl. British India 5: 23, 1886; Blat. & Hallb. Fl Ind. Desert 26(4): 969, 1920; Bhandari Fl. Ind. Desert 299, 1990; Parmar in Shetty & Singh, Fl. Rajasthan. 2: 745, 1991.

Local name: Phog, Phogro

Fig. 7.1. Occurrence of *Calligonum polygonoides* at Barmer.

What is Phog or *Calligonum polygonoides*? Phog is an important component of natural vegetation of the Thar

Desert (Fig. 7.1). A typical sand-dune plant, often covering the entire dune, forming rarely pure associations, but often with *Acacia jacquemontii* Benth. and *Aerva persica* (Burm.f.) Merrill. The young branches which are green and fleshy, appear during July-Sept. This is a drought resistant shrub providing fodder, fuel and food. It is also an excellent binder of shifting sand-dunes.

Fig. 7.2. Flowers and fruits of *C. polygonoides*

Fig. 7.3. Close up of fruits of *C. polygonoides*

How to identify it? A rigid, much branched, almost leafless shrub. Branches terete, pale, glabrous and slender; the internode 3-4 cm long; stipule short, membranous, cupshaped, obliquely truncate and produced upwards at one side. Flowers pinkish, fasciculate in the axils of ocreate (Fig. 7.2 & 7.3); pedicels 1.5-2.5 cm long. Perianth about 3 mm long, the lobes 2 mm long, obovate, cuneate. Ovary tuberculate; styles 4, slightly connate at the base; stigma capitate.

Fruit 6-8 mm long, oblong densely clothed with reddish-brown bristles dilated at base. It flowers profusely and the flowering may start from the year of establishment. The flowering occurs from February to March. The flower buds locally known as 'phogla' are available for a very short period i.e., 1-2 weeks. The fruits mature in the first week of April and most of them fall in the end of April depending upon the climatic conditions. Phog is a prolific seeder and a well grown shrub can produce 1.5 to 3 kg of fruits (Bhandari, 1990).

Where does it naturally occur? It is widespread from eastern Saudi Arabia to the Thar desert. In India, Phog is distributed throughout western Rajasthan particularly in the extremely arid areas on sand dunes and sandy plains (Fig. 7.4). It is one of the pioneer species, that colonizes freshly deposited sands. Within western Rajasthan it occurs in almost all the districts. It is the dominant constituent of shrub associations viz. *C. polygonoides* + *Clerodendron phlomidis*, *Calligonum polygonoides* + *Haloxylon salicornicum*, *Calligonum polygonoides* + *Acacia jacquemontii*. In the extreme arid regions it is found in mixed vegetation.

Fig. 7.4. Natural stand of Phog on sand-dunes of Bikaner

What are its uses?

- **Food:** Flower buds locally called Phogla after light boiling or frying are used in preparing traditional dish called 'Rayata' by mixing these buds with curd during summer. Young branches are also chewed to quench the thrust.
- **Fodder:** Camel browses young green and preferably fleshy branches. It is also used as fodder for sheep, goat and cattle particularly during scarcity period. Dried leaves and young twigs called 'Lasu' are collected, stored and fed to livestock with other feed materials such as guar phalagati (Cluster bean pod husk) or moth bean husk (*Vigna aconitifolia*).
- **Fuelwood:** It has been the chief source of domestic fuel wood and still the most preferred fuel wood species among villagers. Its roots are excellent fuel and hence used by goldsmiths in kilns for making ornaments.
- **Medicine:** Aqueous decoction of twigs is used for sore gums. Root decoction mixed with "Katha" is used as gargle. The flower buds are supposed to have a cooling effect and are used against sunstroke. Juice of branches is used for eye washing as an antidote when the milky juice of Aak (*Calotropis procera*) accidently gets in eyes. Phog is recently included in 10 medicinal herbs to be used in 'Charak Programme' in arid region for curing the ill effects of high temperature and extreme climatic conditions.
- **Live hedge:** Used as live hedge to demarcate boundaries of agricultural fields and to protect home yards and agricultural crops from domestic as well as wild animals.
- **Social and religious:** Used in worship in 'Gangor' festival in Rajasthan.
- **Mulch:** Spread as mulch during summer in the crop fields.

- **Material for huts:** Used for making the huts and animal sheds by local inhabitants.

What are the threats to its survival?

Observations at the sites of its occurrence and during regeneration led to identify following threats to its survival:

- The digging of roots loosens the soil and accelerates the wind erosion which exposes its roots and kill it.
- It is in high demand as fuel wood in Plaster of Paris industries. So decline in its population is due to large-scale cutting, digging of roots and also changes in the traditional farming system.
- It is restricted in distribution and have species-specific edaphic/climatic requirements.
- Urbanization and irrigated agricultural expansion through IGNP has marginalized it.
- Its heavy exploitation for fuel wood over the years has endangered its natural viable populations in western Rajasthan.
- Its buds are used in 'kadi' to develop taste resulting in very little fruits and seed formation, for regeneration in the next year.

What efforts have been made to conserve it?

- Good quality seeds were collected from one phog dominated site of Jaisalmer and also brought seedlings. Seeds were sown in polybags in the nursery.
- Low germination was observed in field (nursery) condition. Germination started after 15 days of sowing only and maximum was recorded between 22-28 days after sowing.
- Awareness among the farmers have re-established its importance as fodder during drought, sand dune stabilization, soil conservation, use as live hedge around

the field boundaries, its medicinal value as food and for conservation of arid environment.

What can people do to protect it?

- This plant may not be uprooted. Instead it could be lopped so that it regenerates in the next season. Protecting it in farmlands, `orans', around habitations and in forests will save its populations.
- This can be planted on field bunds and field boundaries.

8 *CENCHRUS PRIEURII* VAR. *SCABRA* (POACEAE)

Cenchrus prieurii var. ***scabra*** Bhandari, Fl. Ind. Desert 395, 1978; Pandey in Shetty & Singh, Fl. Rajasthan 3: 995, 1993.

Local name: Lambio-bhurat

What is Lambio-bhurat or *Cenchrus prieurii* var. *scabra*? It is an annual erect grass which is found on sandy habitats. It is threatened and endemic to western Rajasthan (Fig. 8.1). The variety can be distinguished by its inner bristles which are eciliate but scabrous throughout.

Fig. 8.1. Lambio-bhurat on sand-dune of Bikaner district

How to identify it? Annual; Culms erect or geniculately ascending, 30–75 cm long; Leaf-sheaths keeled; Ligule a fringe of hairs; Leaf-blades 8–30 cm long, 5–10 mm wide; Leaf-blade surface scaberulous, rough on both sides. Inflorescence (Fig. 8.2): Peduncle antrorsely scabrous above, spiciform, linear to oblong, 6–14 cm long, 2–4 cm wide; Spikelets subtended by an involucre; Fertile spikelets sessile, 2 in the cluster; Involucre composed of bristles, connate into a disc below (3 mm diam), 15–20 mm long; Involucral bristles deciduous with the fertile spikelets, numerous, inner bristles longer than outer, 15–20 mm long, eciliate but scabrous throughout. Fertile spikelets: Spikelets comprising 1 basal sterile florets, 1 fertile florets, ovate, dorsally compressed, 4–5 mm long, falling entire, deciduous. Glumes: One the lower absent or obscure or two, dissimilar; Lower glume lanceolate or ovate, 2 mm long, 0.5 times length of spikelet, 1-veined, apex obtuse; Upper glume ovate, 3–4 mm long, 0.75 times length of spikelet, 3–5-veined, apex obtuse or acute. Florets: Basal sterile florets barren; Lemma of lower sterile floret ovate, 1 times length of fertile lemma, 5-veined, acute; Fertile lemma ovate, dorsally compressed, 4–5 mm long. 5-veined, margins covering most of palea, apex obtuse or acute. Fruit: Caryopsis ellipsoid, dorsally compressed, 2–2.5 mm long. Its flowering and fruiting occurs during August to December (Fig. 8.3) (Bhandari, 1990).

Fig. 8.2. Lambio-bhurat at peak flowering time

Fig. 8.3. Inflorescence of *C. prieurii* var. *scabra*

How does it differ from *Cenchrus prieurii* var. *prieurii*? *Cenchrus prieurii* var. *scabra* is distinguished by its inner bristles which are eciliate but scabrous throughout (Fig. 8.5). But *Cenchrus prieurii* var. *prieurii* is distinguished by its inner bristles which are densely ciliate towards the base and scabridulous above (Fig. 8.4).

Fig. 8.4. Ciliate inner bristles of var. *prieuri*

Fig. 8.5. Eciliate inner bristles of var. *scabra*

Where does it naturally occur? It is threatened and endemic grass of Indian desert and found only in western Rajasthan (Jodhpur, Barmer and Jaisalmer). It is a typical desert grass found on sand-dunes and sandy-plains (Fig. 8.6).

Fig. 8.6. Natural habitat of *C. prieurii* var. *scabra* on sandy plains

What are it uses?

- This grass is a much liked fodder by all animals.

What are the threats to its survival?

Observations at the sites of its occurrence and during regeneration led to identify following threats to its survival:

- Predation of seeds by goats and sheep.
- Sudden temperature fluctuations kill younger seedlings.
- Habitat loss is major causes of its poor population.

What efforts have been made to conserve it?

Survey of localities from where it has been reported by previous workers was undertaken. It has been found at five sites of Churu district (Bhiyansar, Jalua, Randhisar, Bhadasar, Kohina) and seven sites of Bikaner district (Hansera, Shobhasar, Lakhusar, Tejana, Hada, Udairamsar, Nokha).

Fig. 8.7. Seeds collection from native place

Fruits were collected (Fig. 8.7) from five sites of Churu and Bikaner districts and also brought local soil to sow these fruits in the nursery.

What can people do to protect it

- Protection of existing plants in grazing lands, around habitation and in forests could promote its regeneration.

9 *CEROPEGIA BULBOSA* VAR. *BULBOSA* (ASCLEPIADACEAE)

Ceropegia bulbosa Roxb. var. *bulbosa* Pl. Cor. 1: 11. t. 1795 *et* Fl. Ind. 2: 281, 1832; Wight, Icon. t. 845, 1844-45; Hook.f. Fl. Brit. India, 4: 67, 1883; Parmar in Shetty & Singh, Fl. Rajasthan 2: 477, 1991.

Local name: Hedulo

Fig. 9.1. *Ceropegia bulbosa* var. *bulbosa*

What is Hedulo or *Ceropegia bulbosa* var. *bulbosa*? The genus *Ceropegia* includes more than 200 species in the world. Several species of *Ceropegia* are narrow endemics and fall under one or another IUCN Red-list category (Nayar and Sastry, 1987). Species of *Ceropegia* possess curious flytrap flowers (Percival, 1969) with great diversity in flower design (Fig. 9.1); corolla size, shape and coloring pattern; corona structures and mechanisms for illumination of essential organs (Yadav, 1996).

Fig. 9.2. Leaf of *C. bulbosa* var. *bulbosa*

Fig. 9.4. Fruits of *C. bulbosa* var. *bulbosa*

Fig. 9.3. Flowers of *C. bulbosa* var. *bulbosa*

How to identify it? A small, perennial, twining herb, from a spherical oblong, turnip-like root-tuber (3 × 2 cm), with 3-5 roots from its base; Stems annual, very slender, terete, glabrous; Leaves 2-4 × 1.5-2.5 cm, elliptic-oblong or obovate (Fig. 9.2), lowest almost orbicular, largest towards the middle of the stem, apiculate, rounded or truncate to subcordate at base; lateral nerves 3-5 pairs; petioles 3-10 mm, glabrous, channeled above; Flowers on pedunculate, umbellate cymes (Fig. 9.3); the peduncles 2-3 cm long, arising from between the petioles; pedicels 3-5 mm long, slender; bracts 2 mm long, linear, lanceolate, acute, glabrous; Calyx more or less divided to the base; sepals 2 mm long, lanceolate, acute, glabrous; Corolla 1.5-2.5 cm long; tube much inflated at base, narrowed above, slightly curved, shortly funnel-shaped below lobes, grayish and glabrous outside purplish within and sparingly hairy in narrower portion of the tube; lobes 4-7 mm long, linear from a triangular base, obtuse, folded on their back, connate at tips, deep-purple, villous within, valvate in bud; Outer corona saucershaped; inner corona of 5, filiform, straight lobes which are 1.5-3 mm long; Gynostegium 2 mm long; pollen masses minute, attached to the carries by short caudicles; Follicles 9-12 cm × 3.5-4.5 mm, cylindric, tapering, towards an acute apex, glabrous; pericarp thin (Fig. 9.4); Seeds 7-8 × 3-3.5 mm, ovate-oblong, flattened with a broad submembranous margin; coma 2.5-3.5 cm long. Flowering and fruiting in July to October (Bhandari, 1990).

How does it differ from *Ceropega bulbosa* var. *lushii*? It has oblong or obovate leaves (Fig. 9.2) but *C. bulbosa* var. *lushii* has elliptic or elliptic-lanceolate leaves (Fig. 9.5).

Fig. 9.5. Elliptic leaf of *C. bulbosa* var. *lushii*

Where does it naturally occur? It is distributed in India, Bangladesh, Pakistan, Saudi Arabia, Oman, N Yemen, Ethiopia, Somalia, Kenya and Tanzania. It is found in drier hilly regions (Fig. 9.6) amongst spiny and thorny bushes, which provide protection from grazing animals (Fig. 9.7). It is a CAM plant (Gaikwad *et al.*, 1989) which explains its wide distribution range in drier parts. The plants usually grow amongst *Euphorbia* bushes on which it climbs.

Fig. 9.6. Location of *Ceropegia bulbosa*

Fig. 9.7. *C. bulbosa* var. *bulbosa* at native sites

What are it uses?

- The paste of seeds is dropped in the ear to cure deafness.
- Decoction of tuber (Fig. 9.8) is taken orally to get rid of urinary bladder stone.
- The tuber is eaten either in cooked or raw form.

Fig. 9.8. Tuber of *C. bulbosa* var. *bulbosa*

What are the threats to its survival? Observations at the sites of its occurrence and during regeneration led to identify following threats to its survival:

- It is restricted in distribution and has species-specific edaphic/ climatic requirements.
- Modification and destruction of the habitats resulted in further fragmentation and decline of populations.
- It has some of the most complicated flowers (Fig. 9.9) with highly specialized pollination mechanisms and species-specific pollinators resulting into low fruit and seed setting.
- Perennation of the tubers during rest of the year is fraught with risk to its survival in nature.
- Its tubers are highly susceptible to microbial decay.

Fig. 9.9. Flowers of *C. bulbosa* var. *bulbosa*

- Its tubers are eaten by cowboys, cattlemen and even wild animals.
- Its aerial plant parts are also attacked and eaten by caterpillars (Fig. 9.10).
- Sudden temperature fluctuations also kill them.
- Most of the seeds are lost along slopes and as such do not reach proper niche for germination and seedling establishment.
- Lack of water in rocky areas affects its survival.

What efforts have been made to conserve it? Survey of its previously reported localities was undertaken. It has been found from two sites of Jalore district (Sunda mata and Jagannatha goshala) and two sites of Jhalawar (Ralawata, Chuna Bhati) district.

Fig. 9.10. Leaves and flowers of *C. bulobsa* var. *bulbosa* attacked and eaten by ants and caterpillars

- Seeds were collected from intact follicle before spliting.
- Seeds were removed from the follicle and shade dried. Totally dried seeds would be stored in cotton bags in dry cool places.
- Seeds were germinated in petri plates (Fig. 9.11) and in poly begs/pots with soil + FYM mixture under shade (Fig. 9.12). It ensured good germination.

Fig. 9.11. Seeds germination of *C. bulbosa* var. *bulbosa* in petri plates

Fig. 9.12. Seed germination of *C. bulbosa* var. *bulbosa* in polybags

- Young seedlings were eaten by squirrels, birds, caterpillars and other micro-organisms. Hence, it should be protected by wire mesh cages. Seedlings were watered on alternate days.
- After three months of sprouting, flowering started and after two months of flowering fruit formation began. In winter, leaf yellowing started and after one month, all leaves and fruit fell. Remaining aerial parts also became dry, leaving only tuber in hibernating condition. During the end of summer season, sprouting in the tubers started followed by re-growth of stem. Its two beds have been developed in Desert Botanical Garden (Fig. 9.13).

Fig. 9.13. New block of *Ceropegia bulbosa* var. *bulbosa* at Lead Garden, CAZRI

What can people do to protect it?

- Cutting of *Euphorbia* bushes should be avoided so that it can grow taking its support.
- Uprooting of its tubers should also be avoided.

10 *CEROPEGIA BULBOSA* VAR. *LUSHII* (ASCLEPIADACEAE)

Ceropegia bulbosa Roxb. var. *lushii* (Grah.) Hook.f. Pl. Cor. 1: 11. t. 1795 et Fl. Ind. 2: 281, 1832; Wight, Icon. t. 845, 1844-45; Hook.f. Fl. Brit. India 4: 67, 1883; Parmar in Shetty & Singh, Fl. Rajasthan 2: 477, 1991.

Local name: Hedulo

Fig. 10.1. *Ceropegia bulbosa* var. *lushii*

What is Hedulo or *Ceropegia bulbosa* var. *lushii*? It is one of the endemic species of genus *Ceropegia* in Rajasthan and in the last four decades, it has been reported from a very few locations (Shetty & Singh, 1991). It could be partly due to its concealed occurrence amongst spiny, gregarious shrubs of *Euphorbia caducifolia, Grewia tenax, Rhus mysorensis* and *Dichrostachys cinerea* (Fig. 10.1). Curious flytrap flowers (Percival, 1969) have immense variation in flower design; corolla size, shape and coloring pattern. Corona structures and mechanisms for illumination of essential organs (Yadav, 1996), make this plant highly attractive.

How to identify it? A small, perennial, twining herb, from a spherical oblong, turnip-like root-tuber (2 × 2 cm), with 2-6 roots from its base; Stems annual, very slender, terete, glabrous; Leaves 4-9 × 1-2 cm, elliptic or elliptic-lanceolate (Fig. 10.2), lowest almost orbicular, largest towards the middle of the stem, apiculate, rounded or truncate to sub-cordate at base; lateral nerves 3-5 pairs; petioles 4-10 mm, glabrous, channeled above; Flowers on pedunculate, umbellate cymes; the peduncles 2-3 cm long, arising from between the petioles; pedicels 2-4 mm long, slender; bracts 1.8 mm long, linear, lanceolate, acute, glabrous; Calyx more or less divided to the base; sepals 1.9-2.1 mm long, lanceolate, acute, glabrous; Corolla 1.5-2.5 cm long; tube much inflated at base, narrowed above, slightly curved (Fig. 10.3), shortly funnel-shaped below lobes, grayish and glabrous outside purplish within and sparingly hairy in narrower portion of the tube; lobes 5-7 mm long, linear from a triangular base, obtuse, folded on their back, connate at tips, deep-purple, villous within, valvate in bud; Outer corona saucershaped; inner corona of 5, filiform, straight lobes which are 1-3 mm long; Gynostegium 2.2 mm long; pollen masses minute, attached to the carries by short caudicles; Follicles 9-14 cm × 3-4 mm, cylindric, tapering, towards an acute apex, glabrous; pericarp thin; Seeds 7-12 × 2-2.5 mm, ovate-oblong, flattened with a broad submembranous margin; coma 2.5-3.5 cm long. Its flowering and fruiting occur during July to November (Bhandari, 1990).

Fig. 10.2. Leaves of *C. bulbosa* var. *lushii*

Fig. 10.3. Bud & Flowers of *C. bulbosa* var. *lushii*

How does it differ from *Ceropegia bulbosa* var. *bulbosa*? It is a perennial twining herb having elliptic or elliptic-lanceolate leaves (Fig. 10.2) but *C. bulbosa* var. *bulbosa* is a perennial twining herb having oblong or obovate leaves (Fig. 10.4).

Where does it naturally occur? It is endemic to India. It is found in drier hilly regions amongst spiny and thorny bushes, which provide protection from grazing animals. In Rajasthan, it has been found in Jalore (Sundha Mata Hill area, Jagannath Gaushala area) and in Jhalawar (Khanpur, Chuna Bhati) districts (Fig. 10.5 & 10.6).

Fig. 10.4. Oblong leaf of *C. bulbosa* var. *bulbosa*

Fig. 10.5. Habitat of *Ceropegia bulbosa*

Fig. 10.6. Natural seedlings of *C. bulbosa* var. *lushii*

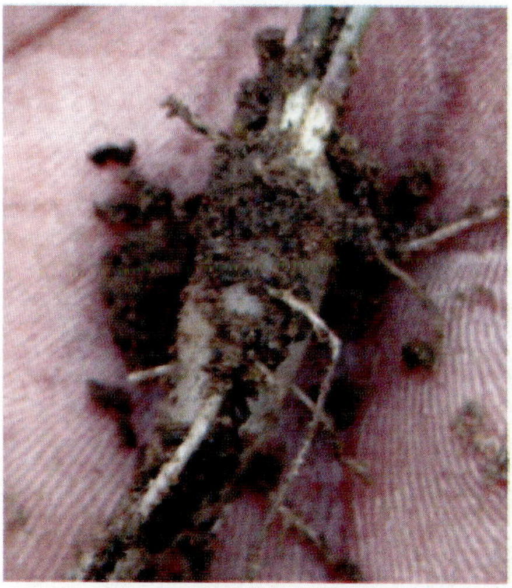

Fig. 10.7. Tuber of *C. bulbosa* var. *lushii*

What are it uses?

- The tuber is eaten either raw or cooked (Fig. 10.7).
- Decoction of tuber is used in urinary diseases.

What are the threats to its survival? Observations at the sites of its occurrence and during regeneration led to identify following threats to its survival:

- It is restricted in distribution and has species-specific edaphic/climatic requirements.
- Modification and destruction of the habitats resulted in further decline and fragmentation of populations.
- It has some of the most complicated flowers with highly specialized pollination mechanisms and species - specific pollinators resulting into low fruit and seed setting.
- Perennation of the tubers during rest of the year faces risks to its survival in nature.
- Its tubers are highly susceptible to microbial infestation.
- Its tubers are eaten by cowboys, cattlemen and even wild animals.

Fig. 10.8. Pupa sucking flowers of *C. bulbosa* var. *lushii*

- Its aerial plant parts are also attacked and eaten by caterpillars.
- Sudden temperature fluctuations also kill them.
- Most of the seeds are lost along slopes and as such do not reach proper niche for germination and seedling establishment.
- Lack of water in rocky areas threatens its survival.

What efforts have been made to conserve it? A survey of its previously reported sites of occurrence was undertaken. It has been found from two sites of Jalore district (Sundha mata and Jagannatha goshala) and two sites of Jhalawar (Ralawata, Chuna Bhati) district.

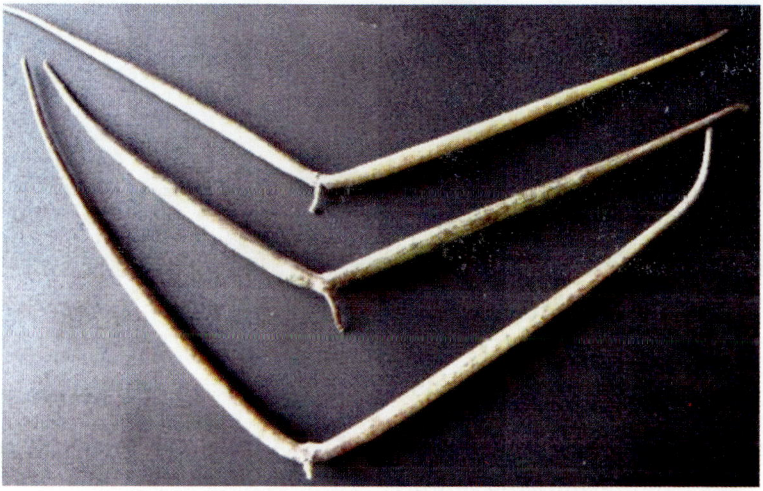

Fig. 10.9. Fruits of *C. bulbosa* var. *lushii*

- Some 3-5 tubers were brought alongwith local soil to propagate it in the nursery of the Desert Botanical Garden at CAZRI, Jodhpur.
- Growth in terms of leaf number, height, number of flower buds, fruits and seeds were periodically recorded till the end of growing period.
- Phenological events were also recorded simultaneously.

- Seeds were collected from intact follicle (Fig. 10.9) before split.
- Seeds were removed from the follicle and shade dried. Totally dried seeds would be stored in cotton bags in dry cool places.
- Seeds were germinated in petriplates in three ways (1) room temperature in light (2) room temperature in dark condition (3) In open sun light condition (Fig. 10.10).

Fig. 10.10. Seed germination of *C. bulbosa* var. *lushii* in petriplates and polybags.

- On emergence of plumule, seedlings were exposed to sunlight and transferred in pots with FYM and soil mixture (1:2) and treated with fungicide (Bavistin) and termiticide (Foret).
- Its roots being very delicate, transplanting in soil required utmost care.
- Its seedlings were watered daily, avoiding water logging.
- Young seedlings were protected by wire mesh cages so that they grew better and stronger. These were regularly irrigated to see its survival and growth.
- Its two new areas are now developed in the Desert Botanical Garden, CAZRI, Jodhpur (Fig. 10.11).

Fig. 10.11. Two new blocks of *Ceropegia bulbosa* var. *lushii*

What can people do to protect it?

- Its tubers may not be removed.
- Its natural habitats should be protected.
- Cutting of woody perennials/bushes should be avoided so that it grows with their support and protection.

11 *CITRULLUS COLOCYNTHIS* (CUCURBITACEAE)

Citrullus colocynthis (Linn.) Schard. in Linnaea 12: 414, 1838; Clarke in Hook.f. Fl. Brit. India 2: 620, 1879; Duthie, Fl. Gangetic Plain 1: 374, 1903; Pandey in Shetty & Singh, Fl. Rajasthan, 1: 333, 1987; Singh & Singh, Biodiversity Desert National Park, 95, 2006; *Cucumis colocynthis* L. Sp. Pl. 1011, 1753.

Local name: Tumba, Tumbo, Indrayan.

Fig. 11.1. *Citrullus colocynthis*

What is tumba or *Citrullus colocynthis*? Genus *Citrullus* belongs to family Cucurbitaceae. This is a large, prostrate, creeping plant with its branches radiating on all sides with a long, perennial tap root. The male flowers are produced not only in large profusion but also throughout the year, while the female flowers are mostly produced after rains (Fig. 11.1).

Fig. 11.2. Tendril of *C. colocynthis*

How to identify it? It is a monoecious herb with perennial root and diffuse or creeping stems which are angled, branched and scabrid. Its tendrils (Fig. 11.2) are simple or bifid, slender and hairy. Its leaves (Fig. 11.3) are variable as 2.5 5 cm in the wild form and 3.8-6.3 cm of the cultivated crop. The leaves are usually deltoid in outline and pale green in colour. Its calyx is hairy, companulate and 5 mm in length. It has female flowers (Fig. 11.3). Its ovary is ellipsoid and densely hairy. Its fruit (Fig. 11.4) is globular, slightly depressed and 5-7.5 cm in diameter. It is variegated green and white. It is glabrous when ripe, filled with a dry spongy bitter pulp and thin epicarp. Its seeds are 4-6 mm long and pale brown in colour. Its fruiting occurs during September to November (Bhandari, 1990).

Fig. 11.3. Leaves and Flowers of *C. colocynthis*

Fig. 11.4. Fruit of *C. colocynthis*

Where does it naturally occur? It is distributed in Spain, Arabia, Africa, West Asia, Sri Lanka, Pakistan, India [Rajasthan (Jodhpur, Barmer, Jaisalmer, Bikaner, Churu), almost throughout]. It either grows wildly (Fig. 11.5) or is often cultivated.

Fig. 11.5. Tumba fruits on sandy plains of Indian desert

What are it uses?

- **Medicinal:** It is used for a wide variety of medicinal purposes: as a purgative, to treat gum disease, to dry up pimples, as a vermifuge, to keep hair from falling out, to relieve pain associated with rheumatism, to treat leucorrhea, as a sexual stimulant in men, to activate the menstrual cycle, and to treat piles and inverted anus.
- **Fruits:** The fruit (Fig. 11.6) is used medicinally for the treatment of various ailments. It is bitter, pungent, cooling, purgative, anthelmintic, antipyretic and carminative. It cures tumours, ascites, leucoderma, ulcers, asthma, bronchitis, urinary discharges, jaundice, enlargement of the spleen, tuberculous glands of the neck, dyspepsia, constipation, anaemias, throat diseases and elephantiasis.

Fig. 11.6. Pulp of *C. colocynthis* fruit

- **Roots:** A paste of its roots is applied to the enlarged abdomens of the children. The cataplasms of its leaves are applied in case of migraine and neuralgia. Its fruit and roots are prescribed in the treatment of snake bite and scorpion sting. Its alkaloidal compounds are prescribed for the treatment of rheumatic pains and stomach disorders.

- **Leaves:** To stop hair from falling, the leaves are pressed and the juice is collected and applied directly to the scalp as a hair lotion. For rheumatism (rehy badi), about 5 g of dried fruit pulp is swallowed with water in the morning and again in the evening, daily. If a larger quantity is taken, it has purgative properties.

What are the threats to its survival?

Observations at the sites of its occurrence and during regeneration led to identify following threats to its survival:

- Its fruits are eaten / exploited for medicinal purposes. Previously, after taking pulp, seeds used to be left in the fields for next year's regeneration. In past 8-10 years the oils extracted from its seeds is finding many uses and hence, seeds are not being left in the field. The older plants due to tractorization are being removed. Thus, new plants are not coming up in nature making it a rare plant.
- Due to presence of waxy coat on the seeds, it shows poor germination in nature.

What efforts have been made to conserve it? Survey of localities from where it has been reported by previous workers was undertaken. It has been found at Barmer, Jodhpur, Bikaner, Jaisalmer, Sriganganagar, Hanumangarh, Sikar, Churu, Jhunjhunun, Pali and Nagaur.

- Seeds were collected from intact fruits before split.
- Seeds were removed from the pulp and shade dried. Totally dried seeds would be stored in cotton bags in dry cool places.
- Seeds were soaked in water overnight and sown in two ways for germination (1) in normal water and (2) seeds kept for few minutes in boiled water and then soaked overnight (Fig. 11.7).

Fig. 11.7. Germination (1-4) of Tumba seeds

- Germination was better in hot water treatment as compared to normal water.
- On emergence of plumule, seedlings were exposed to sunlight and transferred in polypacks with FYM and soil mixture (1:2) and treated with fungicide (Bavistin) and termiticide (Foret).
- Young seedlings were regularly irrigated to see its survival and growth.

- Growth in terms of leaf number, height, number of flower buds, fruits and seeds were periodically recorded till the end of growing period.
- Phenological events were also recorded simultaneously.

What can people do to protect it?

- After harvesting fruits, some of its seeds should be regularly left in the field for enriching its seed bank in soil.
- It should be allowed to grow on field fences, hedge lines and in fallow fields including bare dunes.

12 *COMMIPHORA WIGHTII* (BURSERACEAE)

Commiphora wightii (Arnott) Bhandari in Bull. Bot. Surv. Ind 6: 327, 1964; Parmar in Shetty & Singh, Fl. Rajasthan, 1: 178, 1987; Singh & Singh, Biodiversity Desert National Park, 64, 2006; *Balsamodendron wightii* Arn. In Ann. Nat. Hist. 3: 86, 1839; *B. mukul* Hook.ex Stocks in Hook. Kew Journ. Bot. 1: 259. t.8, 1849; Bennett in Fl. Brit. India, 1: 529, 1874; *Commiphora mukul* Engl. in DC. Monogr. Phan. 4: 12, 1883; Blatt. & Hallb. In J. Bom. Nat. Hist. Soc. 26(1): 232, 1918.

Local Name: Guggal, Guggulu

Fig. 12.1. *Commiphora wightii* in wild (Jalore)

What is Guggal? The Genus *Commiphora* belongs to family Burseraceae and commonly known as guggal (Fig. 12.1). It is medium sized shrub. This species is the source of Guggal or Indian Bdellium. It is largely used in medicines, incense and perfumery. It occurs on rocky hills and piedmonts.

How to identify it? It is a slow growing, much branched, shrubby plant, 2-3 m high with silvery and paper like grayish or grayish-brown bark (Fig. 12.2) peeling off in small pieces, branches spirally ascending spinescent with young parts glandular and pubescent. Leaves 1-3 foliate, leaflets sessile to sub sessile, terminal one largest rhomboid to ovate, margin serrate to dentate (Fig. 12.3). Flowers small, maroon-pinkish. It exists in three forms i.e., female (Fig. 12.4), male (Fig. 12.5) and polygamous having male and bisexual flowers both; petals 4-5, ligulate, brownish red recurved at the tip and four to five times as long as sepals, calyx has glandular hairs forming cylindrical cup, stigma inconspicuously bilobed, stamens 8 (4+4) in male flowers slightly exerted, in female flowers staminodes, fruit drupe (Fig. 12.6), reddish ovate with 2 celled stone, rarely three or four valved and red when ripe, stone covered with four strips of yellow aril, seeds lustrous and brown. It's flowering and fruiting occurs during September to April (Bhandari, 1990).

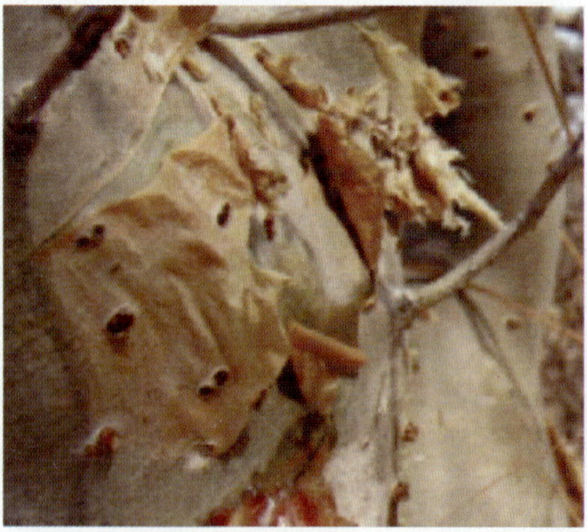

Fig. 12.2. Bark of Guggal

Fig. 12.3. Leaves of Guggal

Fig. 12.4. Female flowers of Guggal

Fig. 12.5. Male flowers of Guggal

Fig. 12.6. Fruits of Guggal

Where does it naturally occur? In Indian sub-continent, *Commiphora* occurs in Pakistan, Baluchistan and India. In India it is found in Rajputana, Bednore (Rajasthan), Khandesh (Maharashtra), Berar (Maharashtra), Mysore (Karnataka), Kachh (Gujarat) and Bellary (Karnataka). Hills and piedmonts are habitats of *C. wightii*. It is associated with tropical *Euphorbia* scrub which is major subtype of desert thorn forest of India (Fig. 12.7).

What are its uses?

- Guggal is used as a remedy for treating or preventing hypercholesterolemia and related cardiovascular diseases. Its extract includes compounds known for their hypolipidaemic properties — the Z- and E- isomers of guggul-sterone and its related guggul-sterols. *C. wightii* has been used as an inactive pharmaceutical ingredient, binding agent, anti-obesity and cholesterol-reducing agent. Gum guggal (Fig. 12.8 & 12.9) is used in incense, lacquers, varnishes, and ointments, as a fixative in perfumes, and in medicine. Therapeutic uses include treatment of nervous diseases, leprosy, muscle spasms, ophthalmia, skin disorders, ulcerative pharyngitis, hypertension, ischemia and urinary disorders.

Fig. 12.7. Wild Guggal at Akal Fossil Park, Jaisalmer

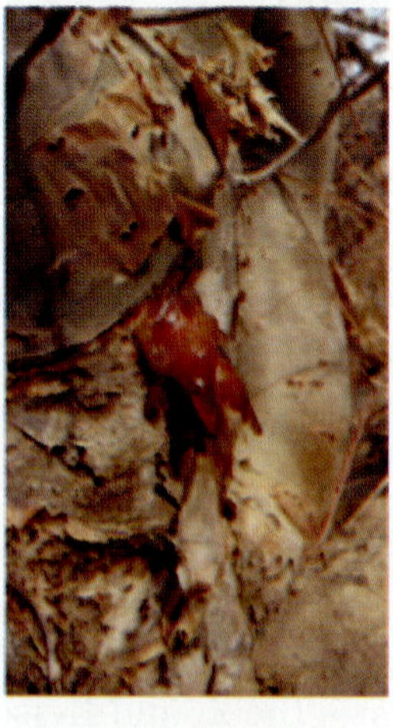

Fig. 12.8. Guggal gum

Fig. 12.9. Guggal gum of trade

- Extract of *C. wightii* is used as pigmenting agent and melanocyte culture agents and used to manufacture pharmaceutical and cosmetic compositions, perfumes. Antioxidant composition is made using *C. wightii* gugulipid components. Cosmetic compounds for skin lightening and anti wrinkle is made from *Commiphora* species. Three percent guggal gum mucilage was found to be a suitable binding agent in tablets and suitable suspending and emulsifying agent for medicines. Traditionally, it is used as an antiseptic on old wounds; as a urine stimulant; on ulcers in the form of lotion. Inhalation of burnt guggal is taken in chronic bronchitis.

What are the threats to its survival? Observation at the sites of its occurrence and during regeneration led to identify following threats to its survival:

1. **Habitat destruction**-Human activities are the principal cause of habitat destruction (Fig. 12.10). Other important causes of habitat destruction include mining and urbanization.

Fig. 12.10. Habitat destruction of Guggal

2. **Overexploitation and faulty tapping techniques** - Oleo-gum resin of the plant is collected by the tribal people using the traditional methods i.e. they give several deep incisions on the stem to get maximum amount of the guggal gum (Fig. 12.11). This crude method increases the amount of guggal gum, but the plant ultimately dies.
3. **Slow growing nature and poor seed germination rate** – It takes about 8-9 years for full maturity to yield gum. The seed germination percentage is often low.

Fig. 12.11. Faulty tapping technique in Guggal

What efforts have been made to conserve it? A survey of its previously reported localities was undertaken. Field surveys to assess potential sites of its occurrence revealed its presence at Jalore (7 sites), Barmer (14 sites), Jaisalmer (15 sites), Sikar (6 sites), Jhunjhunun (1 site), Jodhpur (3 sites), Sirohi (3 sites), Ajmer (11 sites), Rajsamand (2 sites), Pali (7 sites) and Nagaur (1 site) districts.

- Collection of germplasms (Fig. 12.12) was by way of stem cuttings. These stem cuttings were treated with IBA and planted in beds premixed with FYM (Fig. 12.13). Their sprouting behavior in nursery was also studied.

Fig. 12.12. Stem cuttings of Guggal

Fig. 12.13. Guggal stem cuttings planted in beds

- One year old sprouted cuttings of germplasm were transferred to field and recorded monthly observations on growth.
- Seeds were collected in paper packet. These were sown in two ways (1) directly sown and (2) mechanical scarification. Germination percentage was high in seeds given mechanical scarification.
- Studies on germplasm conservation are underway in two blocks at Desert Botanical Garden, CAZRI, Jodhpur.

What can people do to protect it?

- These should not be uprooted for fuel wood or any other purpose.
- Younger plants should not be tapped for gum.
- Tapping needs to be judicious.
- This could be grown in temple yards, sacred groves and under protected environment.

13 *CULLEN PLICATA* (FABACEAE)

Cullen plicata (Delile) Stirton in Bothalia 13: 317, 1981; *Psoralea odorata* Blatt. & Hallb. in J. Bomb. Nat. Hist. Soc. 26(1): 238, 1918; Santapau in J. Bomb. Nat. Hist. Soc. 56(2): 228, 1959; Singh in Shetty & Singh, Fl. Rajasthan, 1: 252, 1987; Singh & Singh, Biodiversity Desert National Park, 80, 2006.

Local name: Jhil.

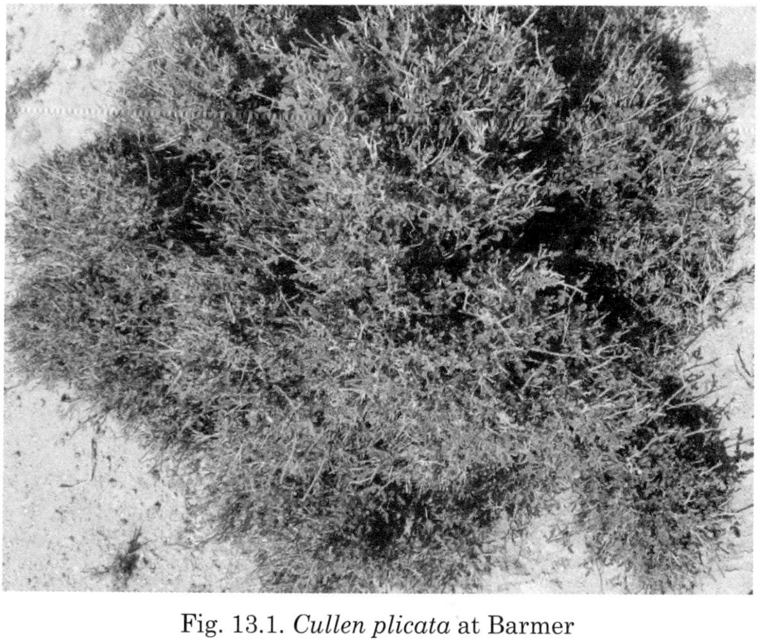

Fig. 13.1. *Cullen plicata* at Barmer

What is Jhil or *Cullen plicata*? *Cullen* Medik is a member of tribe Psoraleeae, all members of which are characterized by delayed development of floral vasculature, the presence of glands and uniovulate ovaries. Jhil or *Cullen plicata* (Fig. 13.1) is the new and revised name of *Psoralea odorata*. Its populations have either declined or disappeared from fresh water bodies called 'Khadins'. Hence, it was selected for its *ex-situ* conservation.

How to identify it? A sweet smelling, perennial herb up to 6 dm high; Stem much branched from base; branches ascending, forming an acute angle with main stem, terete, striate, rough with numerous cicatrices of hardened mucilage, appressedly hairy; Leaves pinnately 3-foliolate (Fig. 13.2); petioles 1-2 cm long, more or less verrucose, grooved; stipules triangular, acute, broad at the base, somewhat falcate, 2-3 mm long, silvery white, strongly nerved, glandular-punctate with a few, golden-yellow, round glands, decurrent forming 3, elevated lines on internodes; lateral elevations becoming suddenly curved; leaflets oblanceolate, apex round or subacute or apiculate; veins prominent beneath, impressed above, appressedly hirsute, thickly hairy particularly at base; terminal leaflet the longest, upto 22 × 6 mm; laterals 12 × 15 mm, all glandular-punctate beneath; margins irregular sinuate, base acute; petiolules 0.5 mm long; Flowers in axillary spikes which are upto 10 mm long (Fig. 13.3); rachis prominently veined, solitary or 2-4 together; pedicels upto 1 mm long, hirsute; bracts minute, ovate, acute, sparsely hairy; Calyx 3 mm long, ribbed and densely covered with white, silvery hairs on the outside; tube 1.5 mm long; lobes ovate, triangular, acute; lowest the biggest; Corolla slightly excerted, 5 × 4 mm; standard broadly obovate, emarginated; margin sinuate; wings oblique rounded, white with blue apex; Stamens connate at base; Ovary compressed, glabrous; style filiform; compressed, curved inwards; stigma minute; Pods 4.5 × 2.5 mm, ovate, brown, glandular-verrucose, hirsute with white hairs; Seeds one, adherent to the pericarp smooth. Flowering and fruiting in November to March (Bhandari, 1990).

Fig. 13.2. Tri-foliate leaf of *Cullen plicata*

Fig. 13.3. Close up of flowers *Cullen plicata*

How does it differ from Bavchi or *Psoralea corylifolia*? *Cullen plicata* is a perennial under-shrub having tri-foliate leaves (Fig. 13.2), pods oblong and glabrous but *P. corylifolia* is an annual herb having simple leaves (Fig. 13.4), pods ellipsoid and densely hairy.

Fig. 13.4. Simple leaf of Bavchi

Where does it naturally occur? It occurs on the margin of water bodies that get filled with rain water and are called `Khadins' (Fig. 13.5). It occurs below the canopy of trees and likes partial shade. It cannot withstand inundation for longer duration. In India, it is sporadically found in Delhi, Gujarat, Haryana, Maharashtra, Punjab and Rajasthan (Alwar, Ajmer, Barmer, Bikaner, Churu, Sriganganagar and Jaisalmer). Its occurrence has been reported from Africa, Algeria, Djibouti, Egypt, Ethiopia, Libya, Mauritania, Morocco, Somalia, Sudan and western Sahara.

What are the threats to its survival? Observations at the sites of its occurrence and during regeneration led to identify following threats to its survival.

Fig. 13.5. *Cullen plicata* on margin of water body

1. It has a very narrow and specialized niche, i.e. on the periodically flooded embanked area of freshwater bodies in drylands. These embanked structures are essentially rain water storage structures which are used for winter cropping (Rabi) in western Rajasthan. There are some over 500 such Khadins in Jaisalmer besides equal number of ponds. However, it has been found in some 10-12 such locations. At some sites in Barmer, it is also found in catchment area of ponds/Agors.
2. Rapid changes in ambient temperature, especially in summers, force the plant to adjust to it by increasing transpiration. However, young saplings with shallow roots dry out in the absence of moisture in the already drying top surface.
3. Since these grow around crop fields, most of its saplings that emerge in soil worked out for cropping are weeded by the farmers.
4. Even whole plants are often uprooted to clean embankments.

5. Grazing, despite so many bioactive compounds that may be grazing deterrent, is responsible for its extermination.

6. Young seedlings are eaten by squirrels, peacocks and caterpillars. Hence, it should be protected by wire mesh cages.

What efforts have been made to conserve it? CAZRI has undertaken survey of localities from where it has been reported by previous workers. It has been found at Harnu, Bothiya at Barmer and Sagar at Bikaner.

1. A detailed survey of its population in Indian desert has revealed its declining populations.
2. From these few populations, seeds were brought but they failed to take root.
3. Young saplings along with soil also did not establish.
4. Freshly collected seeds were shade dried and then germinated in partial shade.
5. Seeds were sown in a mixture of soil and Farm Yard Manure (FYM) and it neither can grow in complete shade nor in direct sunlight so it prefers diffuse sun lights (Fig. 13.6).

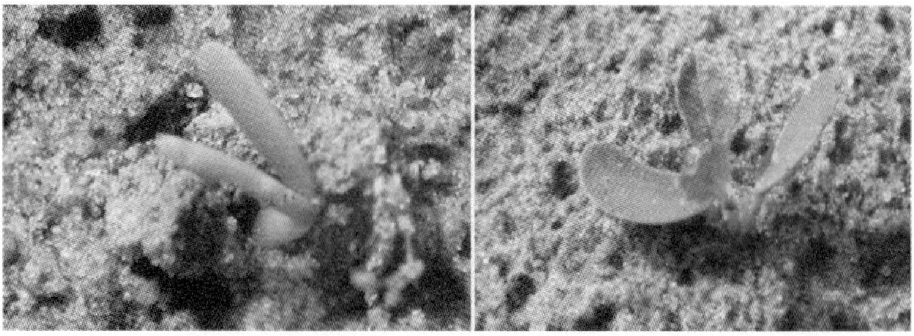

Fig. 13.6. Germination of *Cullen plicata*

6. A new block of *Cullen plicata* has developed in Desert Botanical Garden, CAZRI (Fig. 13.7).
7. It was protected from birds, squirrels, pests and direct sunlight. It was transplanted below the canopy of *Prosopis cineraria*. It has flowered and set

seeds. Hundreds of new plants have now been prepared from these seeds (Fig. 13.8).

8. Watering is provided twice a week.

Fig. 13.7. *Cullen plicata* block in Desert Botanical Garden

Fig. 13.8. Germplasm of *Cullen plicata* in Desert Botanical Garden

What can people do to protect it?

- Anyone removing its branches/leaves/whole plants, could be persuaded not to do so as it will kill the plant. This may be protected in farmland, wastelands, around habitations and in the forests.
- This can be planted in waste lands or grazing-lands. It will provide fodder and fruits and also products of medicinal value.

14 *CYMBOPOGON CITRATUS* (POACEAE)

Cymbopogon citratus (DC) Stapf. in Kew Bull. 1906; Bor, Grass. Burm. Ceyl. Ind. Pak. 128. 1960; Bor, Grass. Burm. Ceyl. Ind. Pak. 126, 1960; Pandey in Shetty & Singh, Fl. Rajasthan 3: 1008, 1993; Bhandari, Fl. Indian Desert 362. 1990; *Andropogon citrates* DC. Cat. Hort. Monsp. 78, 1813; Hook. f. Fl. Brit. India 7: 210, 1896.

Local name: Lemon grass, Nimbu-ghas

What is lemon grass or *Cymbopogon citratus*? The genus *Cymbopogon* belongs to the grass family, *Poaceae* which is a very large cosmopolitan family consisting of about 50 to 60 tribes, 660 genera and 9000 species throughout the world. *Cymbopogon* is a genus of about 55 species of grasses, native to warm temperate and tropical regions of the world. It consists of tall grasses. It is composed of tufted perennial grasses often with aromatic leaves. The inflorescences have two branches with the common stalk not longer than the branches and usually hidden within the spathe. Usually, the inflorescences are very densely clustered in large panicles. It is a rare grass of Indian desert and forms tufts in grasslands (Fig. 14.1). It is a fast growing, lemon scented perennial aromatic grass. Lemon grass can reach a height up to 1.5 m. Lemon grass grows well at a temperature range of 10 to 33°C. It performs well on sandy to clay loam soils with a pH range of 5.0 to 8.4 and of good drainage.

Fig. 14.1. *Cymbopogon citratus*

Fig. 14.2. Inflorescence of *Cymbopogon citratus*

How to identify it? Densely tufted, erect perennial, 1-2 m high. Leaves 20-80 × 1.25-1.75 cm, linear, acuminate, attenuated at the base, glabrous; ligule very short, scarious. Panicles 25-50 cm long (Fig. 14.2), spatheate. Racemes binate, upto 2.5 cm long, purpulish red, villous. Fertile spikelets- 5 mm long (Fig. 14.3), linear-lanceolate, acuminate, glabrous; callus bearded. Glumes subequal, subchartaceous, keeled. Pedicellate spikelet male or neuter, glabrous. Its flowering and fruiting occurs during September to February (Shetty and Singh, 1993).

Fig. 14.3. Close-up of spikelet of *C. citratus*

How does it differ from *Cymbopogon jwarancusa* subsp. *jwarancusa*? Sessile spikelet of *Cymbopogon citratus* is not awned and has lemon aroma in leaves (Fig. 14.3). But in *Cymbopogon jwarancusa* var. *jwarancusa* sessile spikelet is awned and leaves have little aroma (Fig. 14.4).

Where does it naturally occur? *Cymbopogon* is a group of grass species, mostly native to tropical and sub-tropical climates including Africa, China, India, South America and Australasia. Many species grow rapidly in their natural habitats. They also are commercially cultivated. Lemon grass is native to southern India and Ceylon, Indonesia and Malaysia. Lemon grass is widely cultivated in the tropics and subtropics. It is rare plant of Indian desert, found on sandy plains and some time it is cultivated for essential oil extracted from leaves (Fig. 14.5).

Fig. 14.4. Inflorescence of *C. jwarancusa*

Fig. 14.5. Lemon grass in nature

What are its uses?

- This grass is much liked fodder by all animals.
- Citral oil is also used for deodorants, waxes, polishes, detergents, and insecticides where its low cost is attractive.
- It is used as a spice for flavoring.
- It is also used to extract citral or citronella to use as an insect repellent.
- It can be used as a pain-reliever for muscle strains and cramps, an anti-inflammatory for bug and mosquito bites, and is used as an antibacterial and antifungal treatment for various skin disorders.
- The herb is also known to help regulate digestion, eases fevers and headaches, and relieves coughs. It even helps reduce symptoms of depression and can act as a mood-enhancer.
- It has a lemony flavour and can be dried and powdered or used fresh. The stalk itself is too hard to be eaten, except for the soft inner part. However, it can be finely sliced and added to recipes. It may also be bruised and wholly added as this releases the aromatic oils from the juice sacs in the stalk. It is commonly used in teas, soups and curries. It is also suitable for poultry, fish and seafood.
- It is excellent for tired and aching feet. Lemon grass oil may be used in the treatment of acne, athlete's foot, excessive perspiration, flatulence, insect- repellant, muscle aches, oily skin, scabies and stress. Safety precautions should be taken when using *Cymbopogon citratus* in aromatherapy and it should not be used for children.

What are the threats to its survival? Observations at the sites of its occurrence and during regeneration led to identify following threats to its survival:

- Predation of seeds by goats and sheep, besides ants and other arthropods.
- Sudden temperature fluctuations kill younger seedlings.
- High wind velocity causes loss of seeds by sand burial that results in poor germination.

What efforts have been made to conserve it? Survey of localities from where it has been reported by previous workers was undertaken. It has been found only at one site from Chhoti-Undhari in Barmer district.

Fig. 14.6. Lemon grass in Desert Botanical Garden

- Fruits were collected from one site of Barmer district and also brought local soil to grow them in the nursery of the Desert Botanical Garden at CAZRI, Jodhpur.
- The plant was established successfully in medicinal block of Desert Botanical Garden in CAZRI, Jodhpur (Fig. 14.6).

What can people do to protect it

- Protection of existing plants in orans, around habitations and in forests could promote its regeneration.

15 *EPHEDRA CILIATA* (EPHEDRACEAE)

Ephedra ciliata Fisch. & Mey, ex C. A. Meyer, Monogr. Gatt. Eph. 100. 1846; Bobrov in Komarov, Fl. URSS. 1:198. 1934; Pandey in Shetty & Singh, Fl. Rajasthan, 2: 815, 1991.

Synonym: *Ephedra foliata* var. ***ciliata*** Stapf in Akad. bliss. Wien, Math-Naturwiss. Kl. Denkschr. 56: 49.1889; ***Ephedra foliata*** Boiss. & Kotschy ex Boiss. Diagn. ser 1: 7.101.1846; Boiss, Fl. Or. 5:716. 1884; Parker, For. Fl. Punjab. 536.1918; ***Ephedra rollandii*** Maire, Bull. Soc. Hist. Nat. Afrique N. 27: 26.

Local name: Phok, Lana, Suo phogaro

What is *Ephedra ciliata*? *Ephedra* is a genus of gymnosperm shrubs, the only genus in the family Ephedraceae and order Ephedrales (Fig. 15.1). These plants occur in dry climates over a wide area. They are also called Joint-pine, Joint-fir, or Mormon-tea. *Ephedra* is also sometimes called sea grape, although that is also a common name for "Coccoloba uvifera". Only one species *Ephedra ciliata* is present in the Rajasthan. Common throughout the area on sand or gravel or even rocks, gregarious or growing as large climber with thick stem and pendulous. When the male and female plants are growing in close proximity the female plants are profusely laden with numerous milky-transparent fruits, the whole plant presenting an excellent sight. The bracts are very fleshy and completely cover the seeds in the initial stage but as they mature and dry-up, the two bio-convex seeds are exposed.

Fig. 15.1. *Ephedra ciliata* in gregarious habitat

How to identify it? It is a dioecious much branched, climbing shrub. Branches slender, smooth, striate, the older 7-8 cm in diameter, more or less knotted. Leaves 2-3 at each node, shortly connate at the base, 25-30 × 1-1.5 mm, setaceous. Male flowers: spikelets ovate, tetragonous, 1-3 together, on 1-2 cm long peduncle; flowers 6-24; bracts 1.5-2 mm long, as much as broad, round, obtuse, connate upto their length from the apex, ciliate. Stamina-column shortly exserted. Female flowers (Fig. 15.2) in sessile 2-3 flowered cymes or fascicles or on 2-10 mm long peduncles; bracts 3 pairs; outer 1×1 mm long, margins scarious, ciliate, the inner most 5 mm long, connate. Fruit ovoid-globose, 7-8 × 6-7 mm, milky white semi transparent, fleshy and two seeded. Seeds 2-7 × 2-3 mm, plano-convex, brown, glabrous acute at apex, rounded at the base, glabrous and dorsally keeled. It's flowering and fruiting occurs during January to April (Bhandari, 1990).

Where does it naturally occur? It is distributed in Afghanistan, Syria, Pakistan and India. In India, it occurs in Punjab and Rajasthan. In Rajasthan, it is only reported from Jodhpur, Churu, (Fig. 15.3) Barmer and Jaisalmer.

Fig. 15.2. Flowering at winter in *E. ciliata*

Fig. 15.3. *Ephedra* in Churu district

What are it uses?

- The fruits (Fig. 15.4) are eaten by natives in the time of scarcity. It is a substitute of *E. gerardiana.*

Fig. 15.4. Buds and fruits of *E. ciliata*

- Extract of plant has ephedrine which is used in respiratory disorders and heart diseases.
- It is also used in diphtheria and pneumonia. The dried powder or infusion is useful in acute rheumatism.

What are the threats to its survival?

Observations at the sites of its occurrence and during regeneration led to identify following threats to its survival:

- They have restricted distribution and found only in protected habitats.
- Male and female plants are separate; so if male plant is not present nearby female plant, then seed formation does not occur hampering natural recruitment.
- Modification and destruction of the habitats resulted in further decline and fragmentation of populations.
- Its aerial plant parts are affected by insect pests and fungal infections.

What efforts have been made to conserve it? Herbarium specimen and published floras of the region were examined to locate sites of its occurrence. Those localities were surveyed from where it has been reported by previous workers. It has been found only at Kalteda and Laxmi Narayan Dhani in Nagaur district; Viratra mata area in Barmer and a few more places.

- Some 10-20 stem cuttings were brought along with local soil to grow it in the nursery of the Desert Botanical Garden at CAZRI, Jodhpur.
- Seeds sown in polythene bags with soil-FYM mixture, germinated 8 days later but germination was very poor (Fig. 15.5).

Fig. 15.5. Seed germination of *Ephedra ciliata*

- Growth in terms of leaf number, height, number of flower buds were periodically recorded till the end of growing period.

What can people do to protect it?

- It can be protected in farmland, grazing-lands, around habitations and in forests.

16 *LASIURUS SINDICUS* (POACEAE)

Lasiurus sindicus Henr. in Blumea 4:514. 1941; Bor,189; *L. hirsutus sensu Bhandari,* Fl. Ind. Des. 420. 1978. *Non* (Forsk.) Bioss,1859; Pandey in Shetty & Singh, Fl. Rajasthan, 3: 1071, 1993; Singh & Singh, Biodiversity Desert National Park, 181, 2006.

Local Name: Sewan

Fig. 16.1. Flowering plant of *Lasiurus sindicus*

What is Sewan? The Sewan is the dominant perennial tussocky grass of the Indian desert (Fig. 16.1).

How to identify it? A tufted perennial, from a more or less oblique and woody, rhizomatous rootstock with many, extra and intervaginal shoots from the base, often appearing almost shrubby. Culms ascending from branching base, 6-9 dm high, terete, smooth or the lower internodes often slightly rough and more or less hairy rigid. Leaves 7.5-15 cm long, 2.5-4 mm wide, the lower sometimes much longer, linear, finely acuminate, convolute or flat; sheaths terete, glabrous or nearly so, the upper sheaths shorter than the internodes, glaucous or mealy white, the lower crowded, long persistant glabrous or sparingly ciliate upwards; ligule a fringe of hairs. Racemes (Fig. 16.2) solitary, erect, up to 10 cm long, densely villous and more or less fragile; joints 4-5 X 2 mm, shorter than sessile spikelet's, subclavate, densely villous on the back, dorsally much compressed, rather ciliolate along the angles; pedicels keeled along the inner side, 4.5 X 1.2-15 mm, otherwise similar to the joints. Sessile spikelets (Fig. 16.3) at each node of the rachis 1 or 2, imbricate, pale-green. Glumes 4; lower involucral glume 1 cm long, coriaceous, 7-9 nerved, acuminate, unequally 2-toothed at the apex, densely silky villous; upper involucral glume 6 mm long, ovate, aristately acuminate, 3-5 nerved, the acumen ciliate at the tip; lower floral glume hyline, ovate-oblong, acute, 3-nerved, paleate; upper floral glume ovate, acute, glabrous, with incurved margins. Pedicellate spikelet similar to the sessile one, but shorter, with both flowers male or reduced. Its flowering and fruiting normally occurs during October to December and second spell of flowering in March to April (Bhandari, 1990).

Where does it naturally occur? North Africa, Nigeria, Ethiopia, Iraq, Somalia, Egypt, Arabia, Abyssinia, Pakistan (Sindh) and India (N.W. Rajasthan). It prefers sandy plains but also grows on low dunes, hummocks and light-textured alluvium. It grows where annual rainfall is below 350 mm (Fig. 16.4).

Fig. 16.2. Raceme of *L. sindicus*

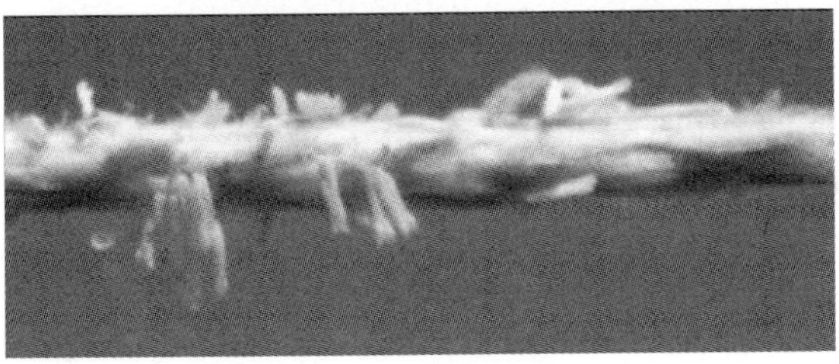

Fig. 16.3. Spikelet of *L. sindicus*

Fig. 16.4. *Lasiurus sindicus* on sandy plains

What are its uses?

- The Sewan grass is much liked fodder by all animals.
- It is highly efficient user of energy and water.
- It is well adapted to desertic conditions and most productive perennial grass. It is rightly called 'King of Desert Grasses'. Sewan grass may be used to stabilize desert sand dunes. It is useful to improve rangelands.

What are the threats to its survival?

Observations at the sites of its occurrence and during regeneration led to identify following threats to its survival:

- Overgrazing of grasslands especially during drought years causes degradation of precious root reserves.
- Due to waterlogging conditions, e.g. IGNP command area, it has been exterminated in larger stretches previously occupied by it.
- It is habitat specific as it grows only on sandy hummocks and dunes.

- During its peak flowering season in October to December, a short spell of rain causes flowers drop and remainder also don't reach maturity.
- Seed damage by desert rodents, ants, spiders, etc. cause loss to its seed bank.
- Its seeds do not germinate during inadequate rains because of inhibitors and consequently imbibed seeds in the absence of moisture die.
- Seeds are often buried under sand where intense insolation and heat often kill them.

What efforts have been made to conserve it? Survey of localities from where it has been reported by previous workers (Barmer, Bikaner, Jodhpur and Sriganganagar) revealed that now it occurs at few sites i.e., Sadasar, Mundsar, Aadsar (Bikaner district); Golghati, Tagawas, Bhaleri, Bhinyasar, Asasar, Bhadasar, Gogatia (Churu district) Barmer, Jaisalmer, Sriganganagar and Nagaur districts.

Fig. 16.5. A block of *Lasiurus sindicus* - Desert Botanical Garden

- Extensive surveys in different districts like Jaisalmer, Barmer, Jalore and some part of Jodhpur had been done to know its potential areas of occurrence.
- A separate area of *Lasiurus sindicus*, has been made in Desert Botanical Garden of CAZRI, Jodhpur (Fig. 16.5).

What can people do to protect it

- Overgrazing needs to be replaced with deferred and rotational grazing.
- Cut and carry system would encourage its growth.
- Its harvesting should be done after it has dispersed seeds so as to augument its soil seed bank.

17 *MELHANIA MAGNIFOLIA* (STERCULIACEAE)

Melhania magnifolia Blatt. & Hallb. in Journ. Bomb. Nat. Hist. Soc. 26(1): 228, 1918; Bhandari, Fl. Indian Desert 71, 1990; Pandey in Shetty & Singh, Fl. Rajasthan 1: 142, 1987.

Fig. 17.1. *Melhania magnifolia* on rocky habitat

What is *Malhania magnifolia*? It is very rare and endemic plant of Rajasthan. It is found in rocky habitats (Fig. 17.1) in association with *Grewia tenax* and *Commiphora wightii*. It is very elegant when flowers are fully open.

How to identify it? A woody, erect, somewhat spreading undershrub. Stem terete, stellately-grey, downy.

Leaves 7-11 × 3-5 cm (Fig. 17.2), ovate-oblong or obtuse or subacute, apex rarely acute, 7- nerved, cordate or rounded at base, margins irregularly crenate or crenate-centate, the upper surface white, wooly-tomentose. Petioles 2-3 cm, stellately grey, downy. Peduncles axillary and terminal, cinereo-tomentose, 4-5 cm long, 1-5 flowered. Pedicels 10-13 mm long; bract 3, 12-13 × 4-5 mm, persistent, cordate, ovate, acuminate. Petals 5, orange 2 ×1.5 cm (Fig. 17.3). stamens 5, 9-10 mm long. Stigma 5, a subglobose; Capsule 1.5 cm in diam. (Fig. 17.4), densely tomentose, calyx persistent. Seeds angular, minutely dark-blistered throughout (Bhandari, 1990).

Fig. 17.2. Leaves of *Melhania magnifolia*

Fig. 17.3. Buds and flowers of *M. magnifolia*

Fig. 17.4. Mature fruits of *M. magnifolia*

Where does it naturally occur? It is endemic to North-west Rajasthan. It is reported from Jodhpur, Barmer, Kota and Jhalawar districts but it was collected from only one place in Ajmer district.

What are the threats to its survival?

Observations at the sites of its occurrence and during regeneration led to identify following threats to its survival:

- It occurs in isolated area and in patches on rocky surface. Its spread is limited by its specific growth requirements.
- Destruction of the habitats due to mining has greatly declined its populations which ultimately perished.
- Fruit production is very low.
- The cotton boll insects attack on immature fruit, lay eggs which hatch to become adult with in the fruits, thus destroying all seeds. Very few viable and mature seeds are left resulting in very poor natural recruitment.

What efforts have been made to conserve it? A survey of its previously reported sites of occurrence was undertaken. It has been found only from Ajmer (Makopura) district in Rajasthan.

Fig. 17.5. Collection of seeds and fruits of *Melhania magnifolia*

- Its seeds and fruits were collected from its native sites (Fig. 17.5).
- Seeds were sown in three ways: (1) directly in polythene bags and (2) presoaked seeds overnight (3) presoaked seeds for two days were sown. Germination was poor.
- A few plants of this species are new conserved in the Desert Botanical Garden, CAZRI (Fig. 17.6).

Fig. 17.6. *Melhania magnifolia* in Desert Botanical Garden

18 *MORINGA CONCANENSIS* (MORINGACEAE)

Moringa concanensis Nimmo (in Gah. Cat. Bomb. Pl. 43, 1989, *nom. nud.*) ex Dalz. & Gibs. Bomb. Fl. 311, 1861; Hook.f. in Fl. Brit. India, 2: 45, 1876; BH 26(1): 235, 1918; Pandey in Shetty & Singh Fl. Rajasthan, 1: 196, 1987.

Local Name: Sarguro, Sarguda

Fig. 18.1. *Moringa concanensis*

What is Sarguda or *Moringa concanensis*? Genus *Moringa* belongs to Moringaceae, a monogeneric family having 34 species. India harbours only three species of Moringa of which 2 species are in Rajasthan (*M. concanensis* and *M. oleifera*). *Moringa concanensis* is a rare plant in Rajasthan (Fig. 18.1). Its name has originated from Konkan region of the Maharashtra (India) as indicated in its specific epithet.

How to identify it? It is a large tree; Stem glabrous; Leaves 2-pinnate, rarely 3-pinnate, upto 5 dm long, primary and secondary rachises thickened at base, articulated with a single gland at each articulation; leaflets 9-15, 1.5-3.4 × 1.2-2.5 cm, rounded at base, apex also rounded or sometimes retuse (Fig. 18.2), rarely oblique sided, darker above, paler beneath petiolules 2.8 mm long, jointed near the apex. Flowers (Fig. 18.3) in dense, divaricate, minutely pubscent panicles, which are 4-5 dm long. Pedicels 8-12 mm long; bracts 1 mm long, caducous. Calyx 8-12 mm long, thinly-tomentose, segments white, oblong, reflexed. Corolla: petals yellowish-white, oblong, spathulate, reddish near the base, upto 1.5 cm long. Stamens 5; staminodes 4-5; filaments hairy at base (Bhandari, 1990).

Fig. 18.2. Leaves of *Moringa concanensis*

Fig. 18.3. Flowers of *M. concanensis*

How does it differ from drumstick or *M. oleifera*? *Moringa concanensis* is a large tree having bipinnate leaves, leaflets 15-30 mm long and petals with red streaks at base and deep yellow colour seed wings (Fig. 18.4). *M. oleifera* is a medium-size tree having tripinnate leaves, leaflets 12-18 mm long and petals yellow or white without red streaks and white colour seeds wings (Fig. 18.4).

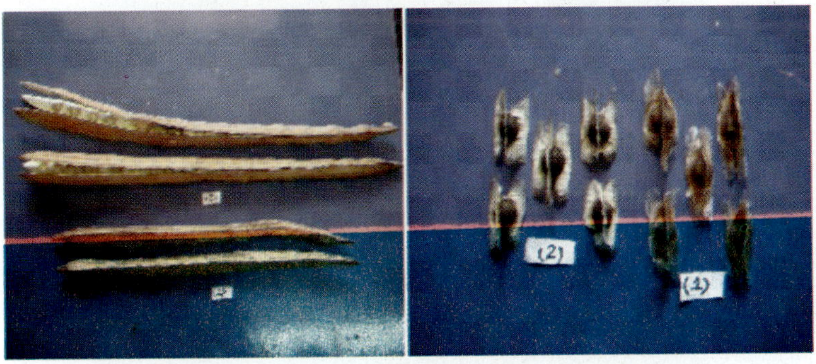

Fig. 18.4. 1- Fruit and seeds of *M. concanensis;*
2 - Fruit and seeds of *M. oleifera*

Where does it naturally occur? A rare plant restricted to hills and rocky uplands. It is distributed in the forests particularly on hills slopes and in wastelands. Almost all *Moringa* species, native to India are now distributed in many tropical and sub-tropical parts of the world like Pakistan, Asia Minor, Africa and Arabia (Mughal *et al.,* 1999). It is a rare plant in Rajasthan, where it has been reported from Dechu, 100 km from Jodhpur, rocky area in Barmer (Bhandari, 1990) and Churu (Gopalpura), Jhalawar (Khanpur), Jaisalmer (Near Kuri), Pali (Piproli) and Tonk (Mayola forest) by Shetty & Singh (1987). Recent survey by CAZRI, Jodhpur revealed this plant at different rocky habitats of Barmer, Jodhpur, Jaisalmer, Nagaur, Pali, Bundi, Jhalawar and Jalore (Fig. 18.5).

Fig. 18.5. *M. concanensis* at Bhopalgarh, Jodhpur

What are it uses?

- **Medicinal:** Entire plant parts (Fig. 18.6) used to cure inflammation, cardiovascular, gastro-intestinal, hematological disorders.

Fig. 18.6. Bark of *M. concanensis*

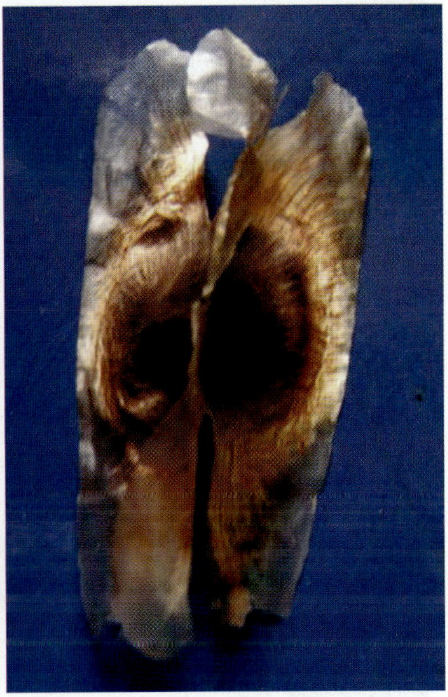

Fig. 18.7. Seeds of *M. concanensis*

- **Edible seed oil:** Seed oil has good oxidative stability during frying (Fig. 18.7). The seed oil is high in oleic acid favoured for replacing poly-unsaturated vegetable oils.
- **Fruits:** The unripe (Fig. 18.8) fruits are eaten by natives.
- **Seed proteins:** The seed protein can be utilized after value addition, used for water purifying and also a good source of non-desiccating oil, commonly known as Ben Oil.

Fig. 18.8. Immature fruits of *M. concanensis*

What are the threats to its survival?

Observations at the sites of its occurrence and during regeneration led to identify following threats to its survival:

- Most of the seeds are lost along slopes and as such do not reach proper niche for germination and seedling establishment.
- Predation of seeds by goats and sheeps (Fig. 18.9).

Fig. 18.9. Predation of seeds by goat

- Modification and destruction of the habitats through mining has declined its populations.
- Sudden temperature fluctuations kill younger seedlings.
- Lack of moisture in rocky areas, does not allow germination and survival of seedlings.

What efforts have been made to conserve it? A survey of its previously reported localities of its occurrence was undertaken. It has been found at oran of Viratra Mata temple in Chauhatan (Barmer), Bhopalgarh (Jodhpur), Tubdi Dhani (Jaisalmer), Degana (Nagaur) and Shivganj (Sirohi).

- Seeds were collected from wild plants in Barmer.
- Freshly collected seeds were shade dried and then germinated in partial shade.
- Seeds were sown in two ways (1) in coco-peat, (2) in a mixture of soil and Farm Yard Manure (FYM).
- Seedling growth was rapid in first 3 weeks in coco-peat medium (Fig. 18.10) while it took four weeks to attain same height in soil medium (Fig. 18.11).

Fig. 18.10. Seedlings of *M. concanensis* in cocopeat

Fig. 18.11. Seedlings of *M. concanensis* in polybags

- Seedling mortality was more in coca-peat than in soil and FYM.
- Young seedlings are eaten by squirrels, birds and peacocks. Hence, it should be protected by wire mesh cages. Irrigation is provided twice a week.
- A new block of *Moringa concanensis* has been raised in Desert Botanical Garden, CAZRI (Fig. 18.12 & 18.13).

Fig. 18.12. New block of *M. concanensis*

Fig. 18.13. *M. concanensis* at Desert Botanical Garden

What can people do to protect it?

- If someone is removing its bark, he could be persuaded not to do it as it will kill the trees. It may be protected in farmland, orans, around habitations and in forests
- This could also be planted on field bunds and field boundaries. It will provide fodder and fruits.

19 *NEURADA PROCUMBENS* (NEURADACEAE)

Neurada procumbens L. Sp. Pl. 441, 1753; Wight, Icon. T. 1596, 1850; Hook.f. Fl. Brit. India 2: 368, 1878; Pandey in Shetty & Singh, Fl. Rajasthan, 1: 308, 1987 Singh & Singh, Biodiversity Desert National Park, 94, 2006.

Local name: Chapri, Chapari.

Fig. 19.1. *Neurada procumbens* in Indian Desert

What is *Neurada procumbens*? The genus *Neurada* includes three taxa: *Neurada procumbens* L. var. *procumbens*,

N. procumbens L. var. *stellata* M. & D. Zohary and *N. aleisawii* Barsotti, Borzatti and Garbari. *Neurada* differs from Rosaceae by its connate carpels within the calyx tube and aberrant pollen. *Neurada* also shows some resemblance to Malvaceae in leaf incisions and verticillate carpels but differs from Malvaceae by having petals inserted on the calyx throat (Wills, 1973). Presently, the genus *Neurada* is usually placed within the distinct family Neuradaceae under the order Rosales (Dahlgren, 1983). Neuradaceae is a small family with 3 genera and 10 species distributed from the Mediterranean region to India and South Africa. In Rajasthan, It is found on sandy and alkaline soils. The seeds germinate, while still in the carpels and the fruits are persistant around the young roots (Fig. 19.1). The collar persists even when the plant is fully grown.

Fig. 19.2. Leaves of *Neurada procumbens*

Fig. 19.3. Flower of *N. procumbens*

Fig. 19.4. Fruit of *N. procumbens*

How to identify it? Diffuse or procumbent, tomentose, densely woody annual herbs commonly found on sand dunes of western Rajasthan; Leaves (Fig. 19.2) are densely hairy on both surfaces, alternate. Flowers white (Fig. 19.3), solitary and born in the axil; Calyx 5, depressed conical about, tube flat, spiny forming conical disc with the ripe carpels, the teeth triangular, 2 mm long; Corolla 5, inserted on the throat of calyx tube, obovate-cuneate or oblanceolate; Stamen 10, inserted on calyx mouth, filaments short; Carpels 10, united with one another and also with the calyx tube, terminated by spinescent style, fruit a depressed cone. Fruit is softly spiny, flat and compressed. The fruit is star shaped (Fig. 19.4), smooth on the underside but has spines on the upper surface that become sharp when dry. The fruit turns hard and woody at maturity. Each fruit has several seeds. It's flowering and fruiting occurs during October to December (Bhandari, 1990).

Where does it naturally occur? It is distributed in North Africa, Arabia, Persia, Afghanistan, Pakistan and India. In Rajasthan it grows in Barmer, Bikaner and Jaisalmer districts. Recent surveys by authors revealed occurrence of this plant at only one sandy habitat of Barmer (Fig. 19.5).

Fig. 19.5. *Neurada procumbens* habitat at Barmer

What are it uses?

- One tea spoon dried powder of the whole plant is given with goat's fresh milk early in the morning to the patient who is suffering from heat stroke during summer season.
- A mouthwash made from root tissue is used to relieve toothache and treat bleeding gums.
- The dried spiny fruit (Fig. 19.6) is crushed in water along with equal quantity of *Corchorus depressus*, poppy seeds (*Papaver somniferum* L.), rose flower

(*Rosa damascena* Miller), almond (*Prunus amygdalis* Batsch) and cardamom seed (*Elettaria cardamomum* (L.) Maton) and candy (Misri) to prepare a cooling agent, locally known as "Thadal" given in summer. This recipe is said to be very effective demulcent and strong tonic especially for male. In winter, the powder of dried fruit of *Neurada procumbens* along with equal quantity of dried fruits is given 1 tea spoon every night with milk as general and nerve tonic for debility and impotency.

- The aqueous extract of *Neurada procumbens* increased the blood pressure of anaesthetized normotensive rats when it was administered orally.

Fig. 19.6. Dry fruits of *N. procumbens*

What are the threats to its survival?

Observations at the sites of its occurrence and during regeneration led to identify following threats to its survival:

- They occur on specific habitats and apparently have specific soil-water requirements.
- Their germination is very poor. Germination rate was just above 1% and death rate was 95% after 4-5 days.

- Modification and destruction of the habitats mainly due to grazing, browsing and cutting have declined its populations.
- Its aerial plant parts are also attacked and eaten by caterpillars.

What efforts have been made to conserve it? A survey of localities of its previously reported occurrence was undertaken. It has been found at only one site from Nakoda in Barmer district.

- Fruits were collected from one site of Barmer and also brought local soil to sow these fruits.
- Fruits were directly used (Fig. 19.7) for germination. Seeds extracted from fruit failed to germinate.
- Presoaked fruits in diffused light and room temperature, resulted in its successful germination (Fig. 19.8).

Fig. 19.7. Fruits of *N. procumbens* used in germination

Fig. 19.8. Seedling of *N. procumbens*

What can people do to protect it?

- It is rarely seen. If spotted, it may be allowed to grow and set fruit.
- While harvesting its fruits, some 30-40% fruits should be left in nature for future regeneration.

20 *SEDDERA LATIFOLIA* (CONVOLVULACEAE)

Seddera latifolia Hochst. & Steud. in Fl. 27. Beil: 8 t. 5 B & C. 1844. Singh in Shetty & Singh, Fl. Rajasthan, 2: 558, 1991; Singh & Singh, Biodiversity Desert National Park, 132, 2006.

Synonym: *Breweria latifolia* Clarke in FBI, 4: 224, 1884; *Bonamia latifolia* (Hochst. & Steud.) Santapau in J. Bomb. Nat. Hist. Soc. 47(2): 342, 1947; *Cressa latifolia* (Hochst. & Stued.) T. Anderson in Jour. Linn. Soc. Bot. 5: Suppl. 1.25, 1860.

Local name: Supdo, Supdio.

What is *Seddera latifolia*? It is a low, bushy, rigid undershrub on open rocky slopes, not very common but forms colonies wherever it occurs (Fig. 20.1).

How to identify it? A low branched, erect herb with slender, woody branches, clothed with dense, velvety pubescence. Argenteo-canescent when young; Leaves 6-13 × 3-8 mm, elliptic-oblong or ovate (Fig. 20.2), obtuse, inconspicuously mucronate, base rounded or acutish clothed with short, adpressed, white hairs above and beneath; petiole when present 1-2 mm long. Flowers solitary in leaf axil or aggregated into short terminal spikes, sessile, bracteoles 2, 2 mm long, linear, lanceolate (Fig. 20.3). Sepals subequal, 4-5 mm long, silky outside like leaves, ovate, tip 2 mm long, acute. Corolla white, equal or slightly exceeding the calyx; limb shortly 5-lobed, tips of midpertaline areas densely hairy.

Stamens equal; filaments dilated at base, with short, rounded appendages. Ovary ovoid, upper portion hirsute; style divided to the base; stigma orbicular. Fruit an ovoid capsule, 4 mm long, light-brown, acute, pubescent near the apex, splitting into 4 valves to near the base. Seeds usually 2, 3 mm long, dark-brown or blakish, glabrous. Flowering and fruiting in Oct. to March (Bhandari, 1990).

Fig. 20.1 *Seddera latifolia* at Jaisalmer

Fig. 20.2. Leaves of *S. latifolia*

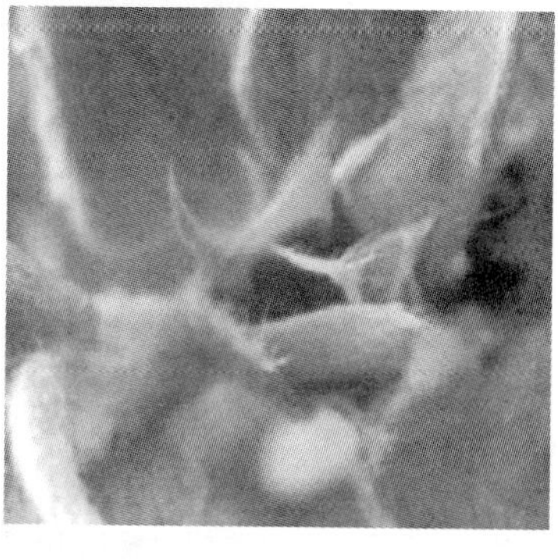

Fig. 20.3. Fruit of *S. latifolia*

Where does it naturally occur? It is restricted to rocky habitats and is true lithophyte (Fig. 20.4). It is distributed in tropical Africa, India. In India it is only present at Jodhpur, Barmer and Jaisalmer districts in Rajasthan. In

Rajasthan it is present near Bada Bagh, Jaisalmer; Kotda, Barmer and in Beriganga and Kailana hills in Jodhpur.

Fig. 20.4. Natural habitat of *S. latifolia* near Jaisalmer.

What are it uses?

- It is used to relieve problems associated with improper milk digestion and gastritis in children under the age of five years.
- Leaves and flowers are dried and ground. About 3/4 teaspoonful of the powder is given to the child each night before bedtime. Treatment continues until symptoms disappear. This drug is particularly useful when the child is vomiting undigested milk and apparently lactose intolerant.

What are the threats to its survival? Observations at the site of its occurrence and during regeneration led to identify following threats to its survival

- It is heavily browsed, being preferred feed of sheep and goat. Its flowers are rarely seen and seed formation is extremely poor.
- Predation of seeds by mites and other arthropods further impoverishes seed bank.
- Premature fruits drop results in poor seed quantity.

- It is restricted in distribution and has a species-specific edaphic/climatic requirements.
- Seeds are also lost along rocky slopes due to water erosion and are lodged in deeper crevices where they fail to germinate.
- Modification and destruction of the habitats mainly due to mining, grazing and cutting has further declined its populations.
- Lack of water in rocky areas adversely affects its survival.

What efforts have been made to conserve it? A survey of its previously reported localities of occurrence was undertaken. It has been found in Jaisalmer, Jodhpur and Barmer districts.

- Seeds were collected from intact follicle before split.
- Seeds were removed from the follicle and shade dried. Totally dried seeds were stored in cotton bags in dry cool places.
- Some 10-15 seedlings were also brought along with local soil to propagate it in the nursery of the Desert Botanical Garden at CAZRI, Jodhpur.
- Seeds were sown in polythene bags which were filled with local soil of natural habitat of *Seddera* plant (Fig. 20.5). Seeds were also sown in a mixture of FYM and soil.
- Germination percentage was higher in native soil (Fig. 20.6) as compared to FYM and soil mixtures.
- Growth in terms of leaf number, height, number of flower buds, fruits and seeds were periodically recorded till the end of growing period.
- Phenological events were also recorded simultaneously.

Fig. 20.5. Germination and seedling growth of *S. latifolia*

Fig. 20.6. Germination of *Seddera* in polybags

- On emergence of plumule, seedlings were exposed to partial sunlight and transferred to polythene pots with FYM and soil mixture (1:2) and treated with fungicide (Bavistin) and termiticide (Foret).
- A new block of *Seddera latifolia* has been developed in Desert Botanical Garden, CAZRI, Jodhpur (Fig. 20.7).

Fig. 20.7. A new block of *Seddera* in Botanical Garden

Fig. 20.8. Germplasm of *Seddera* in green house

- Germplasm of *Seddera latifolia* is now conserved in the green house of the Desert Botanical Garden, CAZRI, Jodhpur (Fig. 20.8).

21 *TECOMELLA UNDULATA* (BIGNONIACEAE)

Tecomella undulata (Sm.) Seem. in Ann. & Mag. Nat. Hist. ser. 3. 10: 30, 1862 et in Seem. Jour. Bot. 1: 18, 1863; Singh in Shetty & Singh, Fl. Rajasthan, 2: 623, 1991; Singh & Singh, Biodiversity Desert National Park, 137, 2006; Blatt. & Hallb. in Bomb. Nat. Hist. Soc., 26(2): 551, 1919; *Bignonia undulata* Sm. Exot, Bot. 1: 35, t. 19, 1805; Roxb. Fl. Ind. 3: 101, 1832; *Tecoma undulata* G. Don, Gen. Syst. 4: 223, 1838; Clarke in Fl. Brit. India, 4: 378, 1883.

Local name: Rohiro, Rohida

Fig. 21.1. *Tecomella undulata* in dense flowering

What is Rohida or *Tecomella undulata*? Commonly known as 'Rohida', the genus *Tecomella* belongs to family Bignoniaceae (Fig. 21.1). It is a medium sized tree and its wood is very useful for furniture. It is found on sandy plains with low hummocks. Flower colour varies from yellow to deep orange. The plants in full flowers give a gorgeous sight.

Fig. 21.2. Different colour flowers of Rohida

How to identify it? It is a medium sized tree. Branches drooping, glabrous, minutely stellately hairy when young. Leaves simple, usually sub-opposite, oblong, apex obtuse, emarginated, margins entire, more or less undulating, slightly cuneate at base; midrib prominent below; petioles terete, glabrous. Inflorescence usually in few-flowered cymes, arranged in racemose manner, terminating short, lateral branches; Calyx campanulate, green or yellow; lobes broadly ovate, obtuse, mucronate often recurved, with few black dots outside. Corolla yellow or orange-yellow, campanulate (Fig. 21.2); lobes 5, subequal. Stamens exserted; filaments glabrous. Ovary on a yellow annular disc; style upto 4 cm long, glabrous; stigma bi-lamillate, elliptic, 3.5 mm long. Fruit: a capsule, curved, linear, oblong, smooth. Seeds winged, wing thin, narrow at the apex of the seed, absent at base. It's flowering and fruiting occurs during January to April (Bhandari, 1990).

Where does it naturally occur? It is distributed in Arabia, Pakistan, India (Maharashtra, Gujarat, Rajasthan). It is an important deciduous, ornamental tree found in the Thar Desert. The natural range of *T. undulata* is restricted to the western parts of India, including southeastern parts of Pakistan. It has been identified as an important agro-forestry tree species in the Thar Desert of western Rajasthan (Fig. 21.3) and southern Haryana, mainly due to its high survival rate even in extreme drought and dry arid conditions. This tree is used for afforestation and landscaping of dry tracts.

Fig. 21.3. *Tecomella undulata* in nature

What are it uses?

- **Wood:** Its wood, known as 'desert teak', is durable and highly valued for engraved furniture, carvings, turnery and toys. It takes a good polish. The heartwood contains a substantial amount of lapachol, which has antifungal and anti-termite properties.
- **Bark:** It is used in indigenous medicines. The bark obtained from the stem is used as a remedy for syphilis, urinary disorders, enlargement of spleen, gonorrhoea, leucoderma and liver diseases.
- **Leaves:** The leaves are eaten by cattle and goats. Biochemical analysis indicated that *T. undulata* leaves have oleanolic acid, ursolic acid and betulinic acid, compounds that are strong HIV prohibitors. Octadimethyl succinate derivatives of oleanolic acid and betulinic acid have been reported to be 24 times more active than AZT, a drug that is currently used for checking the spread of AIDS. Other compounds isolated from the leaves of *T. undulata* are B sitosterol, triacontanol, cirsimaritin, cirilineol, pentatriacontanol and 4,5-dihydroxy-3,6,8-trimethoxy flavone. Both aqueous and alcoholic leaf and stem extracts of *T. undulata* showed growth inhibition of *Salmonella typhi*, a causal organism of typhoid fever.
- **Seeds:** Seeds are used against abscess.

What are the threats to its survival?

Observations at the sites of its occurrence and during regeneration led to identify following threats to its survival:

- Predation of seeds by goats, camel and sheeps.
- It is over exploited plants because of its extensive use as furniture and in fuel.
- When temperature increases sharply, its flowers fall before seed formation, resulting in little regeneration.

- Fruits are blown by high velocity winds during April-May and these are ultimately buried in sands. Such seeds are lost leaving little for future regeneration.

What efforts have been made to conserve it? Survey of localities from where it has been reported by previous workers was undertaken. It has been found from Bikaner (23 sites), Pali (2 sites), Jaisalmer (2 sites), Hanumangarh (2 sites), Sri Ganganagar (3 sites), Ajmer (2 sites), Churu (52 sites), Sikar (6 sites), Jalore (2 sites), Jhunjhunun (6 sites), Nagaur (4 sites), Jodhpur (5 sites) and Barmer (5 sites) districts.

Fig. 21.4. Young plant of Rohida

- Seeds were collected from intact fruits before split.
- Seeds were shade dried and soaked in water for (1) 12 hours and (2) 24 hours.
- Germination was better in seeds soaked in water for 24 hours as compared to 12 hours.

- On emergence of plumule, seedlings were exposed to sunlight and transferred in pots with FYM and soil mixture (1:2) and treated with fungicide (Bovistin) and termiticide (Foret).
- Growth in terms of leaf number, height, number of flower buds, fruits and seeds were periodically recorded till the end of growing period.
- Phenological events were also recorded simultaneously.
- A new block of *Tecomella undulata* was developed in the Desert Botanical Garden, CAZRI, Jodhpur (Fig. 21.4).

What can people do to protect it?

- It may be protected in farmland, orans, around habitations and in forests.
- This could also be planted within the crop lands, field bunds and field boundaries.
- Naturally regenerating saplings in crop lands may be protected to develop into mature plants.

22 *TEPHROSIA FALCIFORMIS* (FABACEAE)

Tephrosia falciformis Ramaswami in Journ. Asiat. Soc. Beng. 12: 125, 1916; Raizada & Sharma in Ind. For. 88(5): 360, 1962; Rolla & Kanodia in Ann. Arid Zone 1(1): 34, 1963; Duthie, Fl. Gangetic Plain 58, 1976; Bhandari, Fl. Indian Desert 129, 1978; Singh in Shetty & Singh, Fl. Rajasthan 1: 263, 1987; Singh & Singh, Biodiversity Desert National Park, 81-82, 2006.

Local name: Rati Biyani.

Fig. 22.1. *Tephrosia falciformis*

What is Rati Biyani or *Tephrosia falciformis*? It is a very rare plant in Rajasthan. It is found on sandy plains (Fig. 22.1) in association with *Tephrosia purpurea, Calotropis procera* and *Prosopis cineraria.*

How to identify it? It is a rigid, perennial, much branched shrub, 3-10 cm high. Branches angular, densely pubescent or adpressed-white-silky. Leaves imparipinnate, 6-13 cm long; stipules 3 mm long, linear, persistent; petioles 2-3 cm long, pulvinous conspicuous; leaflets 5-17, narrowly oblanceolate, cuneate at base; terminal leaflet 3-5 × 0.5-1.5 cm, the lateral leaflet 2.5-4.5 × 0.3-0.8 cm, all argenteo-canescent with adpressed hairs on both surfaces; secondary nerves 11-13 pairs; petiolules of terminal leaflet 3-4 mm, of lateral leaflets 2-2.5 mm. racemes terminal, lax, 10-30 cm long. Flowers usually geminate, one long and the other short pedicellate. Calyx white-silky, tube 2.5 mm long, teeth as long as tube, lanceolate. Corolla purple-red (Fig. 22.2); standard orbicular-cordate 8 × 10 mm, densely silvery-white outside. Style glabrous, slightly incurved; stigma penicellate. Pods 5-10 cm long, falcate or rarely circinate (Fig. 22.3), flat, densely pubescent all over; tip mucronate. Seeds 5 mm long, obscurely reniform, glaucous (Bhandari, 1990).

Fig. 22.2. Flowers of *T. falciformis*

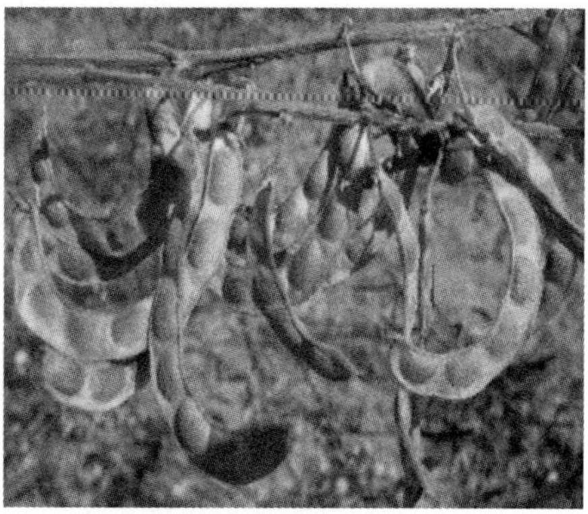

Fig. 22.3. Falcate fruits of *T. falciformis*

How it does differ from *Tephrosia purpurea*? Its pods are falcate, semilunar or circinate and stem base reddish in colour (Table 1). In *Tephrosia purpurea* pods nearly straight or slightly curved towards the tip and stem base is whitish in colour (Table 1).

Table 1. Leaves, stem colour and pods shape of *T. purpurea* and *T. falciformis*

	T. purpurea	***T. falciformis***
Leaf tip		
Stem colour		
Pod shape		

Where does it naturally occur? It is distributed on sandy soils near human habitations. It is reported only from Pakistan and India. In India it is found in Rajasthan (Fig. 22.4) especially from Ajmer and North-West Rajasthan (Jodhpur, Jaisalmer).

Fig. 22.4. *Tephrosia falciformis* on sand-dunes

What are its uses?

- The trunk and branches are used as fuel by local people.
- Its leaves and fruits are used as fodder for cattle.
- It pods are source of a flavanone, falciformin (Khan *et al.*, 1986) and a growth hormone, triacontanol besides longamil, β-sitosterol, lanceolatin-B and lanceolatin-A (Khan *et al.*, 1984).

What are the threats to its survival?

Observations at the sites of its occurrence and during regeneration led to identify following threats to its survival:

- It is restricted in distribution and has species-specific edaphic/climatic requirements.
- Modification and destruction of the habitats due to urbanisation has resulted in fragmentation of its populations which ultimately perished.

- Farmers unknowingly remove it as weed from fallows and croplands.
- Its germination is very poor.

What efforts have been made to conserve it? CAZRI has undertaken survey of localities from where it has been reported by previous workers. It has been found in Nagaur (Barani, Deh), Bikaner (Mukam, Jasrasar, Momasar), Churu (Jaluvu, Khotri, Near Talchapper, Near Churu city, Bhadasar) and Sikar (Kalyanpura) districts.

Fig. 22.5. Seed germination and seedling growth in *T. falciformis*

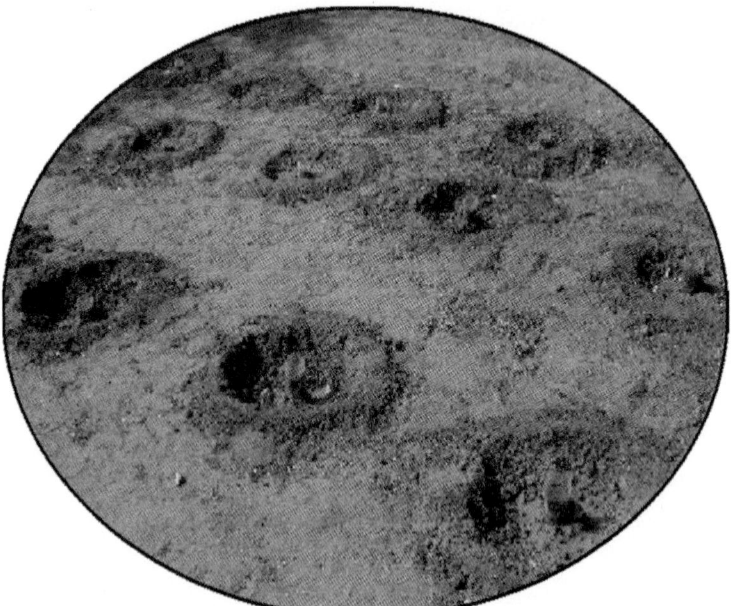

Fig. 22.6. A new block of *Tephrosia falciformis* in Botanical Garden

- Its seeds were collected from its native sites.
- Seeds were sown in two ways: (1) directly in polythene bags and (2) seeds presoaked overnight were sown. The germination was higher in presoaked seeds than in direct sown seeds (Fig. 22.5).
- A separate block of *Tephrosia falciformis* is being developed in the Desert Botanical Garden, CAZRI, Jodhpur (Fig. 22.6).

What can people do to protect it?

- Protection of existing plants in grazinglands, around habitations and forests could promote its regeneration.

23 *ZIZIPHUS TRUNCATA* (RHAMNACEAE)

Ziziphus truncata Blatt. & Hallb. in Journ. Bomb. Nat. Hist. Soc. 26(1): 234, 1918; Pandey in Shetty & Singh Fl. Rajasthan 1: 187, 1987; Bhandari, Fl Indian Desert 93, 1990; Bhandari & Bhansali in Singh *et al.* Fl. India 5: 242, 2000; Singh & Singh Biodiversity Desert National Park Rajasthan 67, 2006.

Local Name: Boti

Fig. 23.1. *Ziziphus truncata* growing on margin of cultivated land

What is Boti or *Ziziphus truncata*? It is found in sandy plains in association with *Calotropis procera* and *Capparis decidua.* It is an endemic plant of Indian desert (Fig. 23.1).

How to identify it? It is a small shrub with divaricate branches (Fig. 23.2); young ones chesnut-coloured, downy. Stipular prickles 2, tomentose at base, other-wise glabrous, up to 1 cm long, one shorter and straight, other longer and recurved (Fig. 23.3). Leaves alternate petiolate, coriaceous, orbiculate, somewhat longer than broad, up to 3.5 cm long, serrulate, subcordate at base, truncate at apex (Fig. 23.4) truncate portion up to 10 mm long and irregularly dentate, glabrous except the puberlous petioles, margins, veins and veinlets on the lower surface, 3-nerved from base; nerves extending right up to apex; midrib with 2-3, equal secondary veins; other 2 veins surrounded by side veins; on exterior; petioles up to 5 mm long. Cymes short, axillary, sessile, tomentose; buds hemispherical, tomentose; pedicels up to 4 mm long. Calyx 5-fid; lobes laterally triangular and opened out, carinate near apex. Petals spathuate, rounded at apex. Disc 10-lobed, with 10-furrows, Style 2, connate, divergent near apex. Fruit globose, 6-8 mm in diam., glabrous, yellow when ripe. Its flowering and fruiting occur during October to November (Bhandari, 1990).

Fig. 23.2. Chesnut-coloured branches of *Z. truncata*

Fig. 23.3. Stipules of *Z. truncata*

Fig. 23.4. Trucate leaves of *Z. truncata*

How does it differ from Bordi or *Ziziphus nummularia*? Its leaves are truncate at the apex (Fig. 23.5) and glabrous on the upper surface and its fruits become yellow (Fig. 23.6) when ripe but *Ziziphus nummularia*'s leaves are rounded at the apex (Fig. 23.7) and more or less tomentose above and its fruits are red when ripe (Fig. 23.8).

Fig. 23.5. Leaves of Boti

Fig. 23.6. Fruits of Boti

Fig. 23.7. Leaves of Bordi

Fig. 23.8. Fruits of Bordi

Where does it naturally occur? This plant is endemic to Indian desert and distributed on sandy hummock impeded below with kankar pan or hard rock in western Rajasthan (Fig. 23.9).

What are it uses?

- The trunk and its branches are used as fuel by local people.
- Its leaves and fruits are used as fodder for cattle.

What are the threats to its survival?

Observations at the sites of its occurrence and during regeneration led to identify following threats to its survival:

- It is restricted in distribution and has specific edaphic/climatic requirements.
- Modification and destruction of the habitats due to urbanisation resulted in fragmentation of its populations which ultimately perished.

Fig. 23.9. *Z. truncata* in open land of Harpalu village in Churu district

- Its over exploitation for fuel and fodder has also diminished its populations.
- Farmers unknowingly remove it as weed from fallows and croplands.
- Its extensive exploitation for commercial purposes has also made it rarer.

What efforts have been made to conserve it? Survey of localities from where it has been reported by previous workers was undertaken.

- It has been found in Nagaur (Alay), Bikaner (Somalsar, Mundsar, Sagar), Jhunjhunun (Khetri Fort, Udaipurwati), Churu (Harpalu, Ojariyo, Near Taranagar, Bhaleri) and Hanumangarh (Bhukarka) districts.

Fig. 23.10. A new block of *Z. truncata* develop in the Desert Botanical Garden.

- Seeds were collected from these sites.
- A separate block of *Ziziphus truncata* has been developed in the Desert Botanical Garden, CAZRI, Jodhpur (Fig. 23.10).

BIBLIOGRAPHY

Balkwill MJ and Balkwill K 1998. A preliminary analysis of distribution pattern in a large, pantropical genus, *Barleria* L. (Acanthaceae). *J. Biogeography* 25: 95 – 110.

Bhandari MM 1990. *Flora of the Indian Desert.* Scientific Publishers (India), Jodhpur

Blatter E and Hallberg F 1918-1921. *The Flora of the Indian Desert (Jodhpur and Jaisalmer).* J. Bom. Nat. Hist. Soc. 26: 218-246, 525-551, 811-818, 968-987; 27: 40-47, 270-279, 506-519.

CAZRI 2007. *CAZRI Perspective Plan Vision-2025.* Central Arid Zone Research Institute, Jodhpur, 94 p

Convention on Biological Diversity, 2011. http://www.iucnredlist.org/documents/ RedListGuidelines.pdf. 15th January, 2011

Dahlgren RMT 1983. General aspects of angiosperm evolution and macro systematics. *Nordic J. Bot.* 3: 119-149.

Duthie JF 1903-29. Flora of the upper Gangetic Plains and of the adjacent Siwalik and Sub-Himalayan tract. Calcutta. Vol. 1-3; Repr. 1952, Vol. 1-2.

Gaikwad DK, Supate AR, Yadav SR and Chavan PD 1989. Occurrence of Crassulacean Acid Metabolism in stem tissues of *Ceropegia juncea* Roxb. *Photosynthetica* 23: 216 - 220.

Hooker J D 1872-1897. *The Flora of British India.* Vol. 1-7. London (Repr. ed. 1954-1961, Kent).

IUCN 2010. www.iucn.org.in guidelines for red list.

Karthikeyan S, Jain SK, Nayar MP and Sanjappa M 1989. *Florae Indicae Enumeratio: Monocotyledonae,* Flora of India, Series, Botanical Survey of India, Kolkata, Vol. 4, 435 p.

Khan HA, Chandrasekharan I and Ghanim A 1984. *Tephrosia falciformis* pods - A new source of Triacontanol. *Annals of Arid Zone,* 23(1): 39-40.

Khan HA, Chandrasekharan I and Ghanim A 1986. Falciformin, A Flavanone from pod of *Tephrosia falciformis. Phytochemistry,* 25(3): 767-768.

Kumar S 1994. Ecology of degraded vegetation and its recovery in the Indian Arid Zone *JIRCAS Int Series* 1: 27-37

Kumar S 1998. Vegetation Resources in the Indian Arid Zone, pp. 89-119 In Faroda A.S. and Singh M. (eds.) *Fifty years of Arid Zone Research in India* CAZRI Jodhpur 478 pp

Kumar S 1999. Biodiversity and its Conservation in Dryland areas. In Singh HP, Ramakrishan YS, Sharma KL and Venkatesh-warlu B (eds.) *Fifty years of dryland agricultural research in India* (CRIDA Hyderabad pp. 57-71

Kumar S 2005. Monsoon Herbage- their Ecology and Potential. In *Proceeding of National Seminar on Conservation, Processing and Utilization of Monsoon Herbage for Augmenting Animal Production.* Indian Society for Sheep and Goat Production and Utilization. P 1-15

Kumar S, Singh G, Parmar PJ, Pandey RP, Sharma SC, Sundermoorthy S, Singh JP, Kasera P, Purohit Chandan Singh and Sen DN 2009. Plant Diversity conservation for sustainability. p. 151-179. In. Kar A, Garg BK, Singh MP and Kathju S (Eds.) *Trends in Arid Zone Research in India.* CAZRI, Jodhpur, 481 p

Linnaeus C 1753. *Species Plantarum.* Vol. **2**. Impensis Laurentii Salvii, Stockholm. pp. 636 – 637.

Mabberley DJ 2008. Mabberley's Plant-Book. A portable dictionary of plants, their classification and uses. Third Edition. Cambridge University Press, Cambridge.

Mughal MH, Ali G, Srivastava PS and Iqbal M 1999. Improvement of drumstick (*Moringa pterygosperma* Gaertn.) – a unique source of food and medicine through tissue culture. *Hamdard Med* 42: 37–42.

Nayar MP and Sastry ARK 1987. *Red Data Book of Indian Plants*, Botanical Survey of India, Calcutta, Vol. 1, p. 367.

Pandey RP, Shetty BV and Malhotra SK 1983. A preliminary census of rare and threatened plants of Rajasthan. In Jain SK and Rao RR (eds.), *An Assessment of Threatened Plants of India* pp. 52-62. BSI, Howrah

Percival MG 1969. *Floral Biology*. Pergaman Press, Oxford. pp.1-243.

Saxena SK 1979. Plant Foods of western Rajasthan. *Man and Environment* 3: 34-43

Saxena SK 1981. Economic plants of Indian arid zone. *Man and Environment* 5: 33-40

Shetty BV and Singh V 1987. *Flora of Rajasthan, Vol. I.* Botanical Survey of India, Howrah.

Shetty BV and Singh V 1991. *Flora of Rajasthan, Vol. II.* Botanical Survey of India, Howrah.

Shetty BV and Singh V 1993. *Flora of Rajasthan, Vol. III.* Botanical Survey of India, Howrah.

Singh JS (Eds.), 2002. National Biodiversity Strategy and Action Plan Natural Terrestrial Ecosystem, BHU Varanasi-422 p

Singh V and Singh M 2006. *Biodiversity of Desert National Park, Rajasthan.* Botanical Survey of India, Howrah.

Willis JC 1973. *A Dictionary of the Flowering Plants and Ferns* 8th Ed. Revised by HK Airyshaw. Camb. Univ. Press, Cambridge.

Yadav SR 1996. Fly trap flowers of the Western Ghats. *Hornbill* 1: 2-7.